IMAGES
of America

PRESCOTT FIRE DEPARTMENT

Prescott Fire Alarm Signals

Courtesy Your MOBILGAS Dealer

Mobilgas

AND BYRL HUBBARD, CONSIGNEE — HI 5-2631

Test	1	Inhalator	4	Outside of City	7
All Clear & 8 o'clock	2	Municipal Airport	5	Whipple Hospital	8
Forest Service	3	Hassayampa Mountain Club	6	National Guard	9

NORTHWEST DIVISION 1 — 1413

- 12 Gurley & Cortez
- 13 Gurley & Montezuma
- 14 Gurley & Granite
- 15 Gurley & McCormick
- 16 Willis & Montezuma
- 17 Willis & Granite
- 18 Willis & McCormick
- 112 Gurley & Summit
- 113 Gurley & Grove
- 114 Gurley & Garden
- 115 Gurley & Willow
- 116 Gurley & Brush
- 117 St. Joseph Academy
- 118 Western & Summit
- 119 Western & Grove
- 121 Western & Garden
- 122 Western & Willow
- 123 Sheldon & Summit
- 124 Sheldon & Grove
- 125 Sheldon & Garden
- 126 Long Beach
- 127 Sonora Ave.
- 128 Garden & Miller Rd.
- 129 Lincoln & Madison
- 131 Robinson Blvd. & Miller Rd.
- 132 Comfort & First
- 133 Dexter School
- 134 Brannen & Fireside
- 135 Brannen & Miller Rd.
- 136 Fourth & Hillside
- 137 First & Hillside
- 138 Lincoln & Hillside
- 141 Ruth & Hillside
- 142 Division & Hillside
- 143 Fair & Miller Rd.
- 144 Fair & Valley
- 145 Fair & Sunset
- 146 Second & Merritt
- 147 Hearthstone & Merritt
- 151 Ruth & Merritt
- 152 Division & Merritt
- 153 Whipple & Third
- 154 Whipple & Hearthstone
- 155 Whipple & Ruth
- 156 Whipple & Division
- 161 Whipple & Miller Rd.
- 162 Whipple & Sunset
- 163
- 164
- 165
- 171
- 172 Hinman & Iron Springs Rd.
- 173 Haining & Iron Springs Rd.
- 174 Iron Springs Rr. & City Limits
- 181 Aainsworth Dr. & Willow Cr. Rd.
- 182 Danita & Northside
- 183 Willow Cr. Rd. & Danita
- 191 Whetstine Ave. & Campbell St.
- 192 Whetstine Ave. & DeMerse Ave.
- 1211 Whetstine Ave. & Northside Dr.
- 1212 Whetstine Ave. & Willow Cr. Rd.
- 1213 Black Dr. & Northside
- 1214 Black Dr. & Willow Cr. Rd.
- 1215 Prescott Heights Dr. & Campbell
- 1216 Prescott Hts. Dr. & DeMerse Ave.
- 1217 Prescott Hts. Dr. & Northside Dr.
- 1218 Prescott Hts. Dr. & Willow Creek
- 1221 Flora St. & Campbell St.
- 1222 Flora St. & DeMerse Ave.
- 1223 Flora St. & Northside Dr.
- 1224 Flora St. & Willow Cr. Rd.
- 1225 Rosser St. & Campbell St.
- 1226 Rosser St. & DeMerse Ave.
- 1227 Rosser St. & Northside Dr.
- 1231 Rosser St. & Oaklawn Dr.
- 1232 Willow Cr. Rd. & Delano
- 1233 Douglas St. & DeMerse
- 1234 Douglas St. & Northside Dr.
- 1235 Oaklawn Dr. & Stevens Dr.
- 1313 Spring St. & Fairgrounds Ave.
- 1322 Dougherty & Fair St.
- 1323 Fairgrounds Ave. & Fair St.
- 1324 The Fairgrounds
- 1325 The Fairgrounds & Robinson Blvd.
- 1326 Dougherty & Center St.
- 1331 Center & West Sts.
- 1332 Keen St. & Mile High Dr.
- 1333 Mile High Dr. & Rose Garden Cove
- 1334 Parr Dr. & Keen St.
- 1335 No. Willow St. & West Sheldon
- 1351 Fairgrounds Ave. & Parr Dr.
- 1352 Norris Rd. & Dougherty
- 1353 Parr Dr. & West St.
- 1361 West St. & Oregon Ave.
- 1362 Linwood Ave. & Manzanita St.
- 1371 Linwood Ave. & Josephine
- 1411 Oregon Ave. & Lindberg Dr.
- 1412 Oregon Ave. & City Limits
- 1413 Pine Lawn Ranch

SOUTHWEST DIVISION 2 — 2131

- 21 Beach & McCormick St.
- 22 Goodwin & Montezuma St.
- 23 Goodwin & Granite
- 24 Goodwin & McCormick
- 25 Carleton & Montezuma
- 26 Sr. High School
- 27 Pioneer Home
- 28 Aubrey & Montezume
- 29 Aubrey & Granite
- 211 Leroux & Cortez
- 212 Leroux & Montezuma
- 213 Leroux & Granite
- 214 Walker & Montezume
- 215 Walker & Granite
- 216 Granite & White Spar Rd.
- 217 Hope St. & Highway 89
- 218 Cherry St. & Highway 89
- 219 Joseph St. & City Limits
- 221 Goodwin & Summit
- 222 Beach & Summit
- 223 Park Ave. & Coronado
- 224 Alto & Willow
- 225 Cory & Frontier
- 226 Heap & Grace
- 227 Smoki & Outwest
- 228 Cory & Sharlot
- 231 Coronado & First
- 232 Hill & Park
- 233 Glenwood & Park
- 234 Seal Ave. & Mistletoe
- 235 Country Club & Park
- 236 Country Club & Vista
- 237 Country Club & Highland
- 241 Carson & Fairview
- 242 Bertrand & Franklin
- 243 Fairview & Perry
- 244 Leroux & Park
- 245 Aubrey & Hassayampa
- 246 Pioneer Dr. & Leroux
- 251 So. of Pioneer Dr. & Leroux
- 252 Ellenwood & Sierra Dr.
- 253 Highway 89 & Copper Basin Rd.
- 254 Park & Copper Basin Rd.
- 255 Middlebrook & Conifer
- 261 Evergreen & Highland
- 262 Highland & Copper Basin Rd.
- 263
- 264
- 271 Linden Rd. & Copper Basin Rd.
- 272 Paul Dr. & Dixon Dr.
- 273 Hassayampa Ave. & Hoffman Dr.
- 281 Hassayampa Ave. & Cop'r Basin R.
- 282 Copper Basin Rd. & Walnut
- 291 Cherry Dr. & Linden
- 2111 Walnut & Linden
- 2112 Linden & City Limits
- 2113 Middlebrook & Aspen
- 2114 Middlebrook & Cherry
- 2115 South Indian Hill
- 2116 North Indian Hill
- 2117 Sunset Park
- 2121 West Gurley & Cory Ave.
- 2122 West Gurley St. & Grace Ave.
- 2123 West Gurley & Josephine
- 2124 Thumb Butte Rd. & City Limits
- 2127 Idylwild Dr. & Lindberg
- 2131 Idylwild Dr. & City Limits

SOUTHEAST DIVISION 3 — 331

- 31 Union & Alarcon Sts.
- 32 Goodwin & Cortez
- 33 Goodwin & Marina
- 34 Goodwin & Alarcon
- 35 Cortez & Carlton
- 36 Cortez & Aubrey
- 37 Aubrey & Marina
- 38 Community Hospital
- 39 Cortez Circle
- 311 Goodwin & Pleasant
- 312 Goodwin & Mt. Vernon
- 314 Goodwin & Washington
- 315 Goodwin & Arizona
- 316 Goodwin & Penn.
- 317 Goodwin & Rush
- 318 Carleton & Marina
- 321 Carleton & Alarcon
- 322 Carleton & Pleasant
- 323 Carleton & Mt. Vernon
- 324 Carleton & Virginia
- 325 Carleton & Arizona
- 326 Aubrey & Alarcon
- 327 Aubrey & Pleasant
- 331 Senator Drive-In

NORTHEAST DIVISION 4 — 4121

- 41 Gurley & Marina
- 42 Gurley & Alarcon
- 43 Cortez & Willis
- 44 Willis & Marina
- 45 Willis & Alarcon
- 46 Cortez & Sheldon
- 47 Sheldon & Marina
- 48 Sheldon & Alarcon
- 411 Gurley & Pleasant
- 412 Gurley & Mt. Vernon
- 413 Gurley & Virginia
- 414 Gurley & Washington
- 415 Gurley & Arizona
- 416 Gurley & Penn.
- 417 Gurley & Rush
- 421 Gurley & Joslin
- 422 Stetson & Overstreet
- 423 Overstreet & Robinson Dr.
- 424 Willis & Pleasant
- 425 Willis & Mt. Vernon
- 426 Willis & Virginia
- 431 Willis & Washington
- 432 Willis & Arizona
- 433 Sheldon & Pleasant
- 434 Sheldon & Mt. Vernon
- 435 Sheldon & Virginia
- 441 Sheldon & Washington
- 442 Sheldon & Arizona
- 443 Sheldon & Rush
- 444 State Highway Shop
- 451 Gurley & Robinson Dr.
- 452 Moeller & Pleasant
- 453 Moeller & Mt. Vernon
- 461 Moeller & Virginia
- 462 Moeller & Washington
- 471 Moeller & Arizona
- 4111 Moeller & Rush
- 4112 Churchill & Washington
- 4113 Churchill & Arizona
- 4114 State St. & Santa Fe Shops
- 4115 Fifth & Navajo
- 4116 Sixth & Navajo
- 4121 Fifth & Merritt

This card showed the locations of all the Gamewell alarm boxes in the city. The system was first installed in Prescott in 1909 with only 20 alarm boxes dispersed across the city. There were over 200 boxes by the time the system was retired in the late 1970s. The boxes were sold in an auction. A few remain as souvenirs in the stations. (Author's collection.)

On the Cover: The 1921 Velie hose-and-chemical wagon arrived in 1922. It lacked a pump, but had a large extinguisher that was agitated by turning a crank, which can be seen just behind the driver. The charged slurry flowed into a hose reel to be discharged. The reel in this picture still needs a hose. (Courtesy of Sharlot Hall Museum, #fi159p.)

IMAGES
of America

PRESCOTT FIRE DEPARTMENT

Eric Conrad Jackson

ARCADIA
PUBLISHING

ISBN 978-1-4671-3178-0

Published by Arcadia Publishing
Charleston, South Carolina

Printed in the United States of America

Library of Congress Control Number: 2013954325

For all general information, please contact Arcadia Publishing:
Telephone 843-853-2070
Fax 843-853-0044
E-mail sales@arcadiapublishing.com
For customer service and orders:
Toll-Free 1-888-313-2665

Visit us on the Internet at www.arcadiapublishing.com

To my late great-grandfather Dr. Robert Mac Stump; I hope my being a paramedic firefighter touched the number of lives you had in mind.

Contents

ACKNOWLEDGMENTS

When I tested for the Prescott Fire Department in 2001, I was asked in my interview, "Why do you want to be a firefighter?" My answer was, "I don't want to be a firefighter. I want to be a *Prescott* firefighter." I went on to explain how I had grown up in this community and that I wanted to give back to it. I further recognized Prescott Fire Department for being the oldest in the state of Arizona, with firefighters who embrace change yet honor the traditions and legacies laid before them.

I have to thank the Prescott firefighters who came before me, who did their part in securing our lineage. Chief Robert Connell Jr. wrote down some of our earliest tales. In the 1950s, Prescott volunteer firefighter Ken Shake also ran a photography business. The department reaped the rewards of having its own department photographer, much as we do today in photographer/firefighter Wade Ward. The late 1970s saw dispatcher Dale Fluck keeping thorough scrapbooks on the department. In the 1980s through to my hiring, firefighter-turned-battalion-chief Tim McElwee kept prolific scrapbooks. These men did the department a huge service in holding on to our heritage.

Of our local organizations, I am grateful to the folks at Sharlot Hall Museum. Their support in capturing our earliest years was paramount. Additionally, the folks at the Prescott Antique Auto Club have to be commended for their efforts to preserve the 1931 Seagrave. My hat is off to them for their work and their excellent documentation of the arrival, departure, and subsequent rebirth of this unique apparatus. I thoroughly enjoyed the Tuesday morning socials.

Lastly, my thanks go to my parents, Jack and Jenise Jackson, for bringing me to this community. I am happy that my kids, Tyler and Madison, have had the opportunity to grow up in a community that is large enough to afford growth opportunities, yet small enough to let you see someone you know every time you go to the grocery store. Keep that in mind kids, you've got 80 uncles watching your every move . . .

Images appear courtesy of the author unless otherwise noted.

INTRODUCTION

Any Prescott Fire Department firefighter can tell you that they work for the oldest fire department in the state of Arizona, and that the department itself started in 1885. Lesser known are the details that led up to its formation.

In 1878, Morris Goldwater had just relocated his business from the southeast corner of Cortez and Goodwin Streets to the corner of Cortez and Union Streets when the dressmaker's shop next door caught fire. His business suffered only minor damage thanks to the rapid response of a bunch of locals who happened to be socializing nearby. It was decided at that time to form the fire brigade, which became known as the Prescott Volunteers. The volunteers were limited to using available water in nearby wells to meet any firefighting needs. Occasionally, well-intentioned miners would use dynamite to remove buildings in the path of an offending fire in order to remove fuels, with mixed results and no doubt to the great dismay of the property owner. Later, then-mayor Goldwater saw to it that wells with hand pumps were installed on each of the four corners of the courthouse square.

In the years 1883 and 1884, there were several major conflagrations in Prescott. On July 5, 1883, a fire destroyed over half of the buildings on the west side of the courthouse. This should not be confused with the "Great Fire," which would not occur until 1900. In February 1884, a noteworthy fire with a fatality occurred at the Sherman House, something of a hotel/boarding house. The Sherman House was located on the south side of the courthouse near where the chamber of commerce building currently stands. What made this fire different was that it claimed the life of S.N. Holmes, owner of the *Jounal-Miner* newspaper. His death spurred the remaining staff of the newspaper to write articles calling for the establishment of a fire department as well as water works. An article from February 15, 1884, stated: "Not only are water works almost a necessity, but an efficient fire department should be immediately organized. No matter what the cost the situation demands it. It is false economy to keep down taxes and welcome the wholesale destruction of property and possibly the frequent loss of life." The recurrent efforts of the staff of the *Journal-Miner* seem to have fallen on fertile soil, as a water system was established by the end of 1884. A reservoir was built on the hill at the south end of Mount Vernon Street, with lines running to the downtown and a steam-pumping plant built on the south side of Miller Creek near North Granite and West Sheldon Streets. With the establishment of a water distribution system from which to work, the official establishment of the Prescott Volunteer Fire Department followed with city ordinance number 25 on March 3, 1885.

Staffing for the initial organization was on a volunteer basis and was divided into four 25-man companies: the Toughs, the Dudes, the O.K.s, and the Mechanics Hook and Ladders. The first paid firefighters were hired in 1906, when the city purchased a horse-drawn wagon laden with hose and ladders. Two engineers were needed to manage the wagon and horses. An additional three engineers would be hired, but not until 1954. A paid chief's position in the form of Chief Jack Custer would follow in 1956. Eventually, in 1975, the volunteer component of Prescott

Volunteer Fire Department would be shuttered, and the department would be strictly known as Prescott Fire Department.

Over the years the City of Prescott annexed surrounding areas, resulting in the department expanding its stations to meet the needs of its citizens. The addition of the airport brought about a station there. Expansion north resulted in stations on Willow Creek and Iron Springs Roads. Annexation of Yavapai Hills meant there was a need to build a station along Highway 69. The city's farther expansion north resulted in closing the Willow Creek barracks in favor of a building farther north on Smoketree Lane. Similarly, expansion south would prompt the closure of the Goodwin Street station in favor of a station farther south on White Spar Road.

Along with the changing needs in coverage came a change in duties and responsibilities. Like many fire departments around the nation and world, firefighters in Prescott became first-line responders in medical responses, first as basic emergency medical technicians, then with the addition of intermediate emergency medical technicians, and eventually adding paramedics to their rosters.

An ability to mitigate all manner of emergencies pressured the department to expand its service role. Prescott firefighters had a need to become skilled at fighting aircraft fires with the addition of Earnest Love Airport, which also necessitated the acquisition of an airport "crash" truck. The local topography, from the cliff bluffs of the Granite Dells and Thumb Butte to the open water of Watson and Willow Lakes, warranted the development of a technical rescue team. A hazardous materials team was created to counter the hazardous chemicals and situations presented by an increasingly industrialized society.

Chief Jack Custer was a strong proponent of prevention, firmly believing that eliminating the hazards in the first place would generate a strong dividend in reducing fire losses in the community. During his 18 years at the helm, Prescott Fire Department saw a strong prevention department developed as firefighters conducted not only business inspections for fire hazards, but expanded their inspections into the homes of the residents of Prescott. The expanded inspection program garnered results too, as one year saw the number of fires fall from 104 to a mere 21.

Another chief with a long-term vision was Darrell Willis. He recognized the ever-growing wildland-urban interface issue faced by Prescott firefighters as more and more heavily wooded areas were annexed into the city. Firefighters over the decades prior to his term had faced multiple wildland fires in and adjacent to Prescott and the problem was getting worse. In response, Prescott Fire Department expanded its wildland firefighter training to meet that hazard. Training would include sending units abroad to fight wildland fires in other states so the firefighters could return to Prescott with an even greater skill set. Chief Willis saw the opportunity to develop a fuels management crew in 2001, which could remove an accumulating hazard in a proactive fashion. This same crew would eventually develop and expand to the capacity of operating as an Interagency Hotshot Crew, the only such municipality-backed hotshot crew in the nation. Tragically, this same crew would perish on June 30, 2013, in the Yarnell Hill Fire, resulting in a loss of 19 lives, the greatest single loss of wildland firefighters in US history.

The current Prescott Fire Department has some 80 firefighters between its structure and wildland divisions. The department operates out of five fire stations, an administration station, and a wildland operations building. These firefighters respond to calls in Prescott's 40-square-mile area in an all-hazard capacity in an effort to meet the needs of a truly diverse community.

One

Arizona's First Department

With the formal establishment of the Prescott Volunteer Fire Department on March 3, 1885, men quickly created four companies to cover the firefighting needs of the community. These four companies were each comprised of 25 men. Fiscal support was provided by the City of Prescott to the tune of $60 per company annually. Additional monies were acquired through donations and fundraisers, which would often net over a thousand dollars.

Toughs Hose Company No. 1 was officially established on April 2, 1885, and was comprised largely of the saloon owners and patrons. The city purchased a two-wheel hose cart and housed it near the center of Whiskey Row. The hose cart had 600 feet of two-and-a-half-inch hose. As these folks had a late-night lifestyle and the saloons never closed, this company had a distinctly better response time when fires started at night.

The Dudes Hose Company No. 2 was comprised of the more respectable residents of Prescott. Amongst its ranks were bankers, clerks, and storeowners. Their hose cart was housed midway down Elks Hill on the north side of the street and shared quarters with the Mechanics. Being halfway up Elks Hill gave the Dudes an advantage in quickly reaching fires downhill.

The Mechanics Hook and Ladder Company No. 1 was formed of what today would be termed "construction workers," being mostly carpenters, painters, and the like. They shared quarters with the Dudes, but did not operate a hose cart. Rather, they had a lightweight, hand-drawn, four-wheeled cart. Mounted on the cart were an assortment of ladders and plaster hooks.

The O.K. Hose Company No. 3 started later and was not recognized by the city until August 22, 1888. It was manned by livery stablemen and teamsters. They were configured differently than the other teams in that they had a four-wheeled cart like the Mechanics, but it was loaded with 600 feet of two-and-a-half-inch hose like those found on the hose carts. The O.K.s were housed on the northeast corner of Fluery Avenue and Gurley Street.

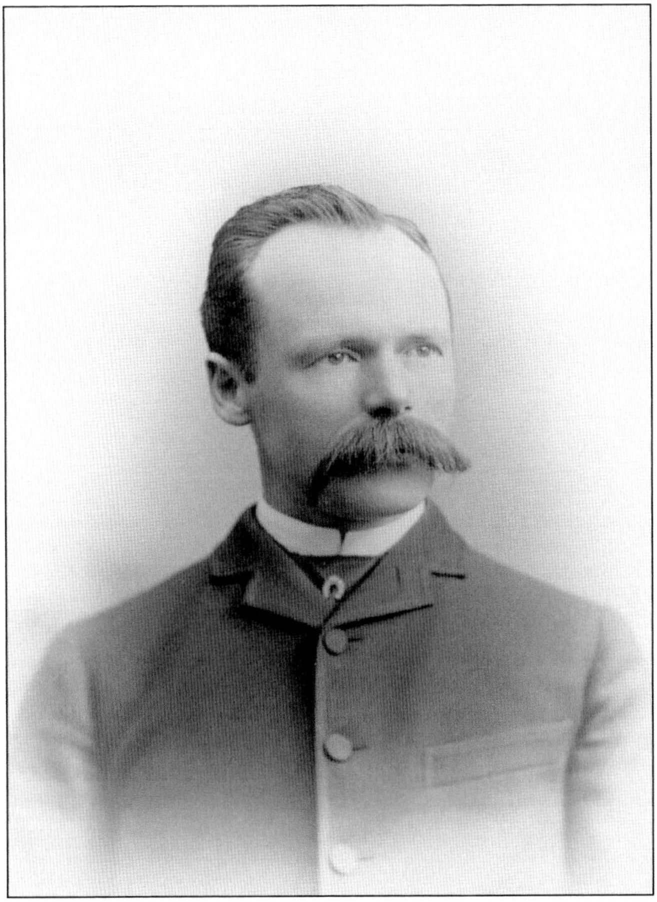

Morris Goldwater was instrumental in the creation of the first firefighter affiliations in Prescott. The threat of fire to his business operations in 1878 served as motivation for creation of the first team, the Dudes. Later, as mayor of Prescott, Morris Goldwater would see to the installation of a water system in the downtown district and the formalized establishment of a city-supported fire department. While Goldwater never served as chief, he remained an active member of the fire department. He is seen in the photograph below pulling an antique hose cart in the Armistice Parade of 1931, over 50 years after he started the department. (Both, courtesy of Sharlot Hall Museum: left, #po0792pc; below, #pa102p.)

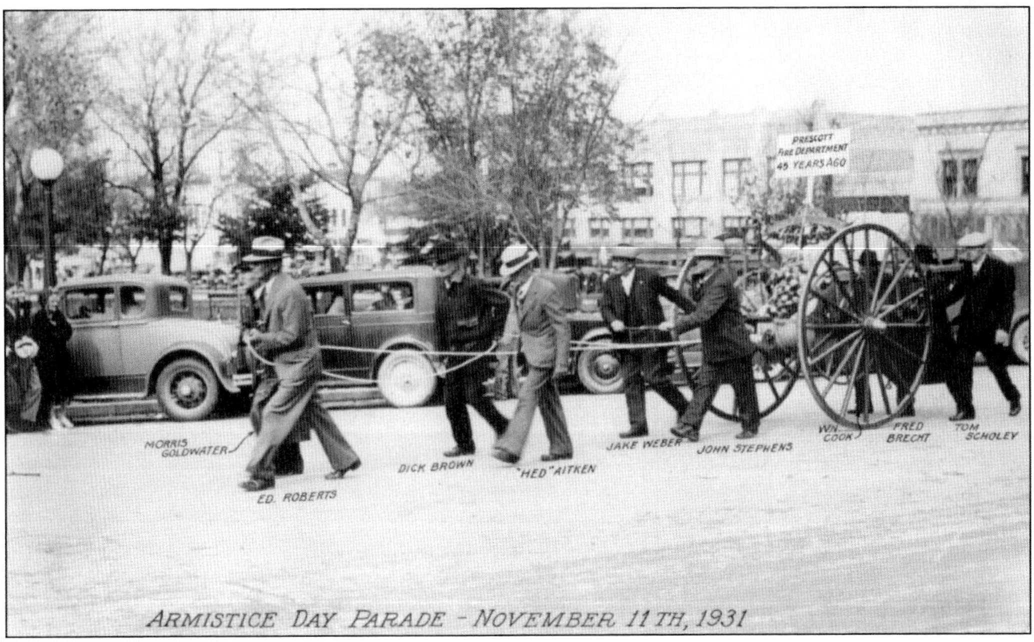

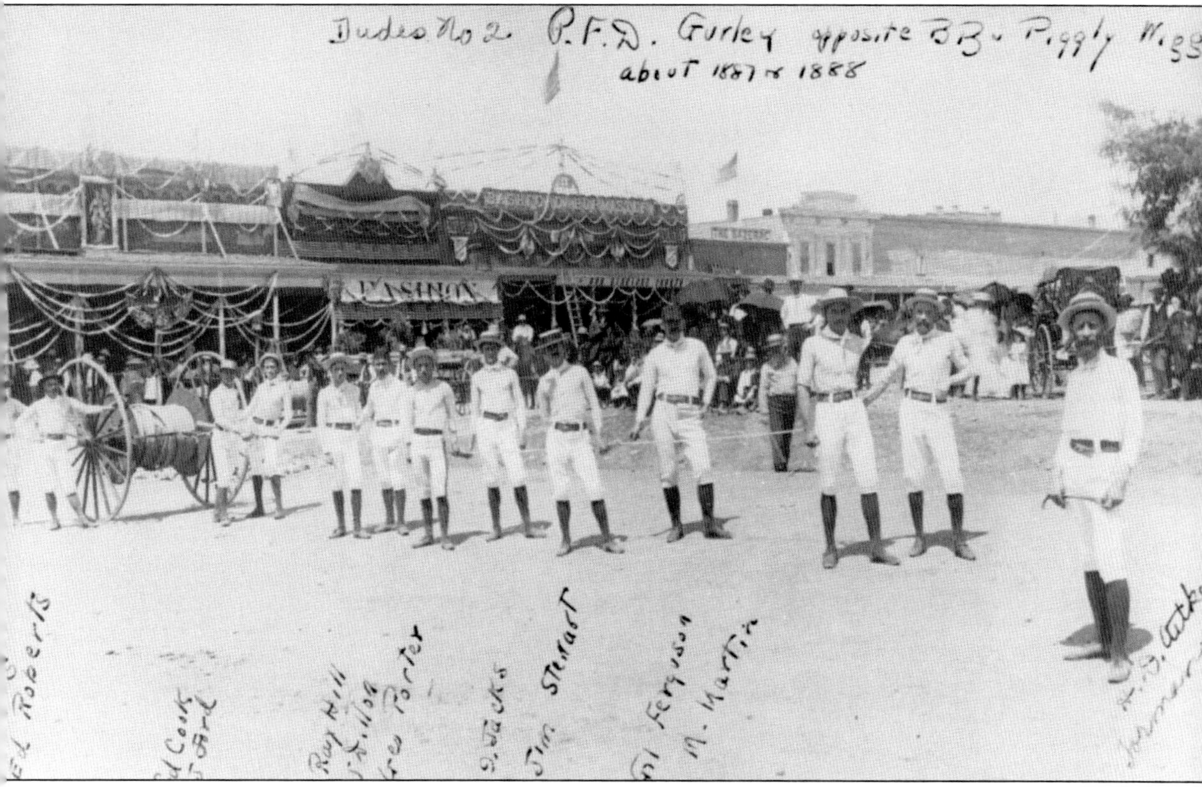

With Gurley Street decorated in celebration of Independence Day, the Dudes No. 2 ready themselves for one of the earliest races in what would become an over-100-year tradition of competition. While some documents claim that the Toughs No. 1 were formed before the Dudes, fire records clearly indicate that the Dudes No. 2 and the Mechanics No. 1 both responded on March 31, 1885, to a fire at George Curry's home, Curry being a member of the Dudes himself. This predates the Toughs organization. At the end of 1885, the Toughs argued that fires predating their establishment should not be counted in the year's tally. By only counting fires from April 2 forward, several responses of the Dudes were effectively removed. The Toughs proudly posted in bold in the newspaper, "We Won the Medal." (Courtesy of Sharlot Hall Museum, #fi102p.)

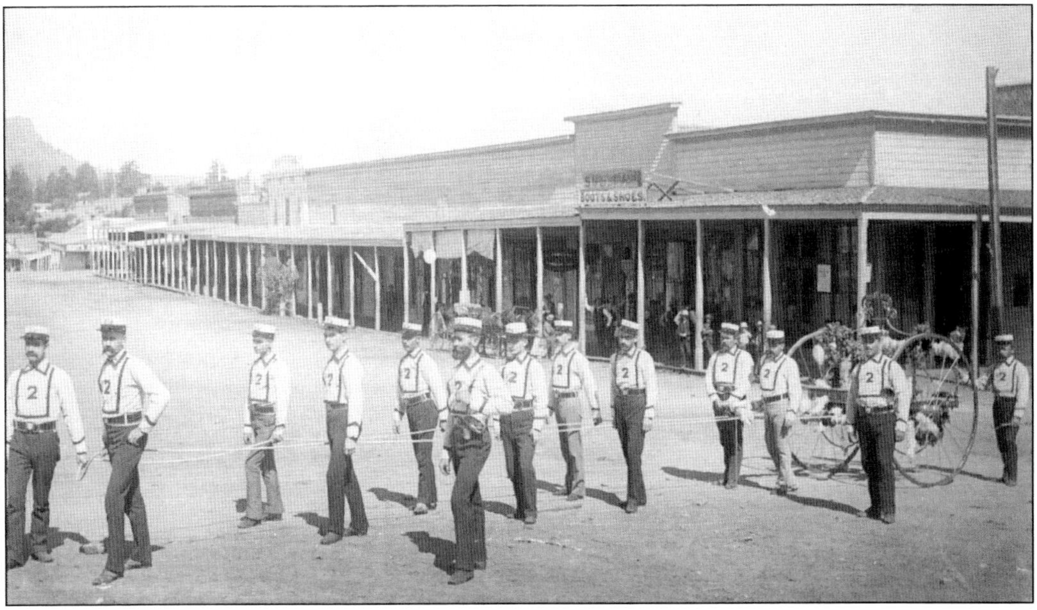

Seen here are the early dress uniforms of the Dudes. The above 1887 photograph shows Hedrick Aitken still sporting his beard. Their hose cart has been amended with name plaques and a lantern rack, also showing their name. The 1891 image below shows their new Class A uniforms as they prepare to participate in the Fourth of July parade. The blur in the background are the Hooks as they move into the parade line. An additional hose cart can be seen in the background, still in service as it is not parade ready; it may be the hose cart that was purchased in July 1881, prior to the formal organization of the department. (Both, courtesy of Sharlot Hall Museum: above, #fi140p; below, #fi115p.)

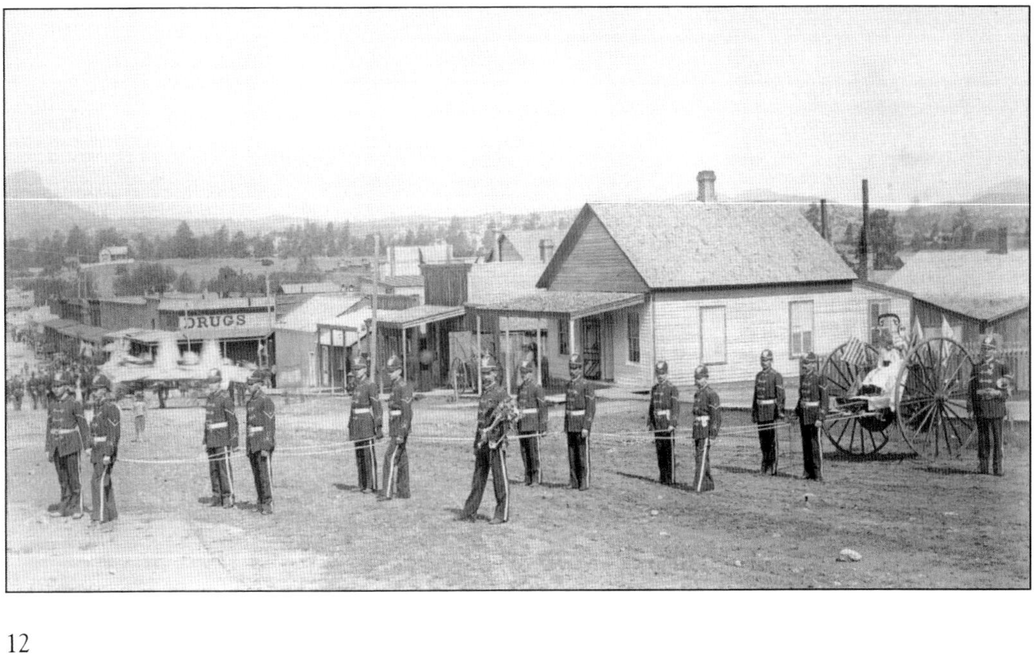

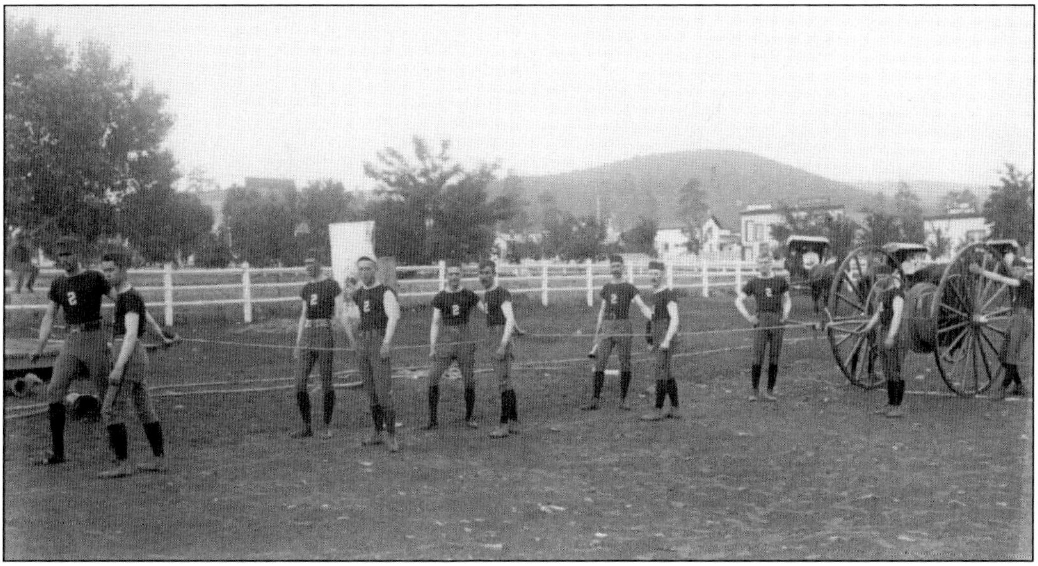

Above, the Dudes racing team is seen in July 1891. Hedrick Aitken (holding the nozzle) had shaved off his characteristic beard and simply sported the classic firefighter's mustache. Aitken would take the reigns as chief from 1895 to 1897. He was active in the local entertainment scene, performing in the local opera houses, and was also the vice president of the Bashford-Burmister store. In the photograph below are, from left to right, (first row) J.M. Aitken, A.W. Robinson, F.H. Waite, H.D. Aitken, J.P. Dillon, and W.E. Bayless; (second row) H.D. Derwin, W.B. Jones, R.H. Oliver, G.D. Ferguson, J. Weber, F.A. Tritle, and E.B. Pentland. (Both, courtesy of Sharlot Hall Museum: above, #fi114p; below, #fi143p.)

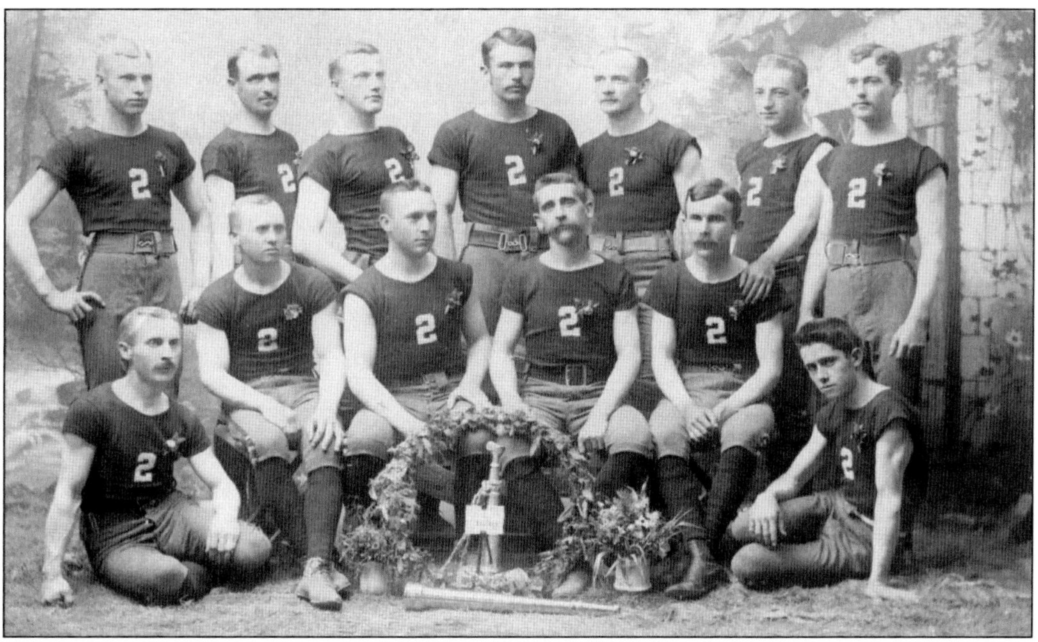

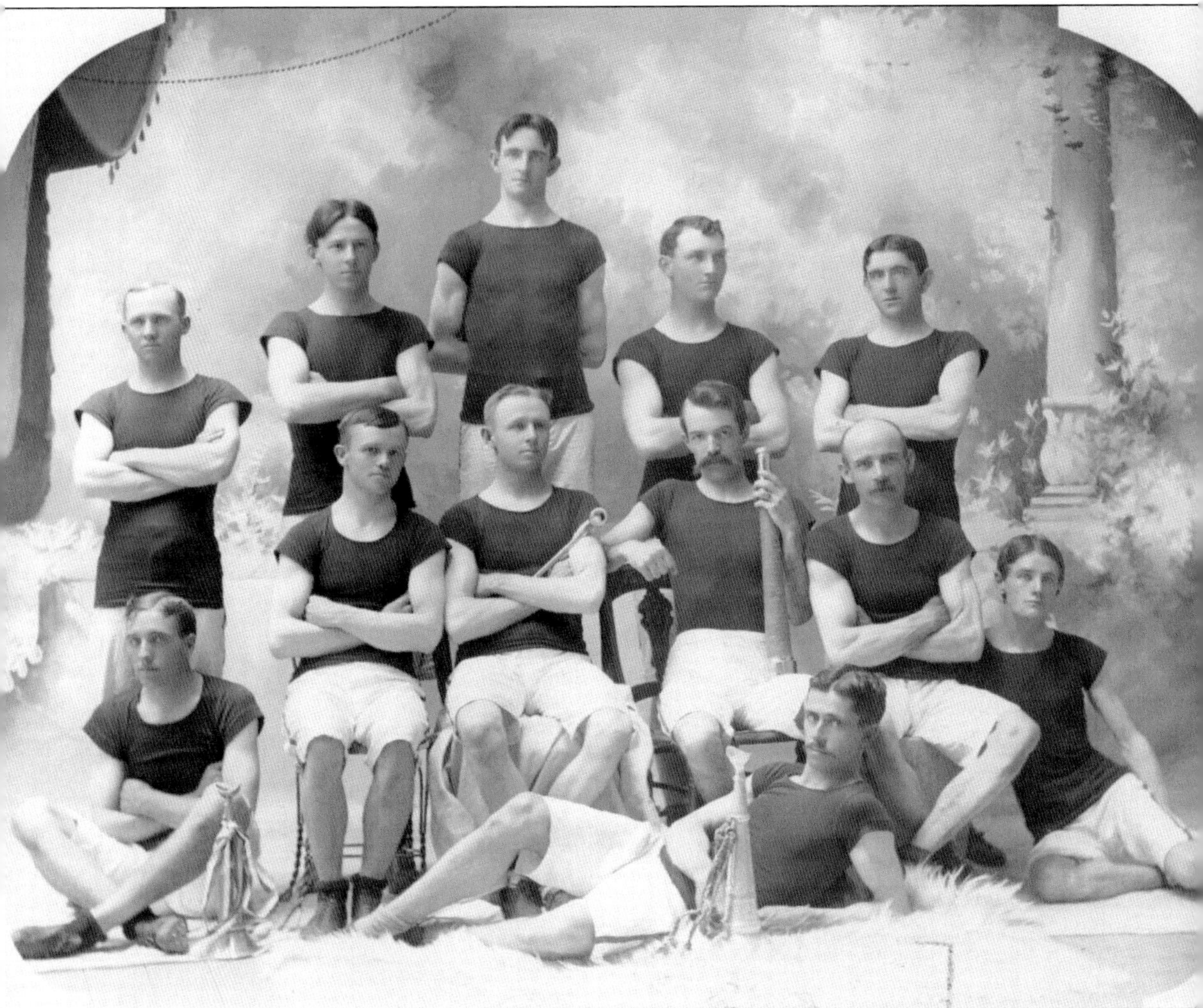

This photograph of the Dudes was taken in 1898, the same year that a slew of arson fires led the city council to appoint six specially chosen "sober and good men" to serve as night watchmen in order to "stop the spirit of incendiarism cropping out of our midst." These watchmen were granted authority to arrest or "wing if necessary" any person out at night under suspicious circumstances. The *Journal-Miner* newspaper staff supported the move, stating, "Taxpayers will never complain, but always approve of such expenditures of public revenue." (Courtesy of Sharlot Hall Museum, #fi146p.)

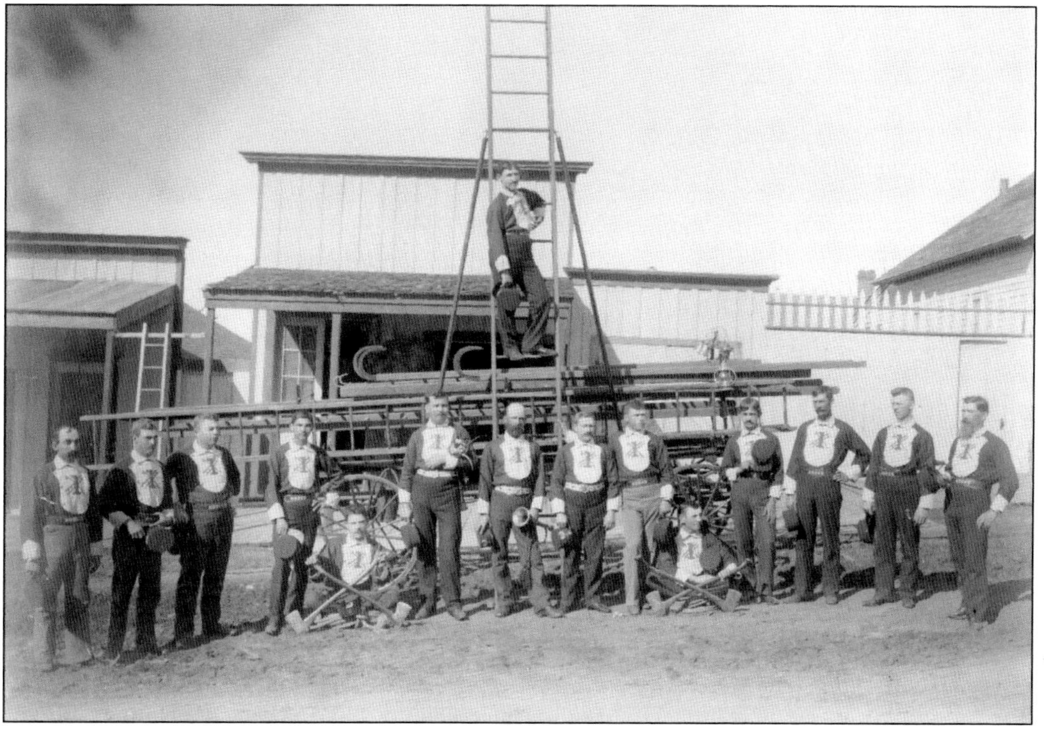

Taken in 1890, this photograph is of the Mechanics Hook and Ladder Company No. 1. On the ladder is Julius Jacoby. Standing below him are, from left to right, Charley Ackers, Joe Engle, Prichett, Tinker, Joseph Tiernan, Nash Burnett, Gene Engle, J.L. Munds, Frank Bevers, unidentified, and Fred Brecht. Unfortunately, the two seated men are also unidentified. Tiernan would be chief by the time of the Great Fire in 1900. (Courtesy of Sharlot Hall Museum, #fi108p.)

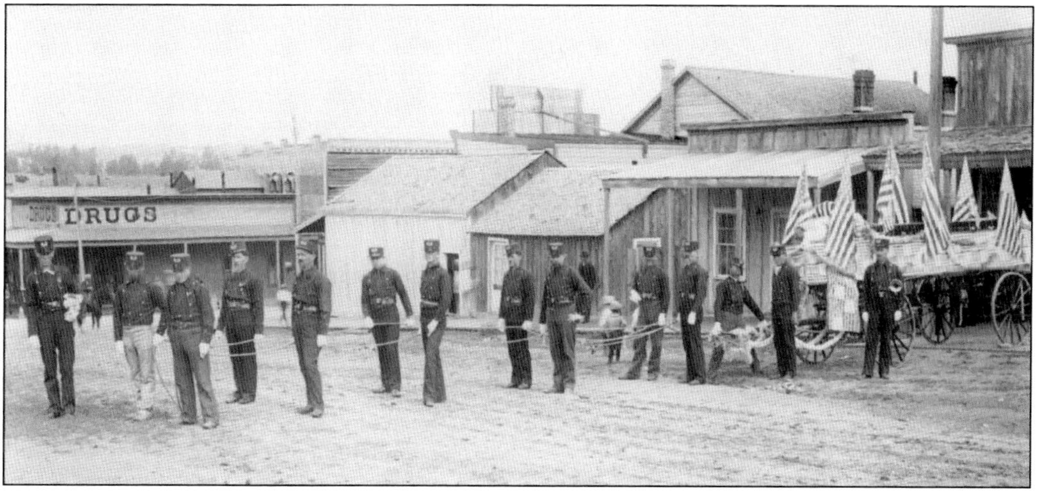

Ladders and hooks, the actual equipment, were commissioned by the council in 1875, well prior to the department's creation. The town marshall saw to their placement around town for use in emergencies. The Mechanics Hook and Ladder crew likely formed in 1879 after the city council requested that "a hook and ladder brigade of volunteer firefighters be organized and supplied as soon as possible with a truck and buckets and the necessary firefighting equipment." (Courtesy of Sharlot Hall Museum, #fi139p.)

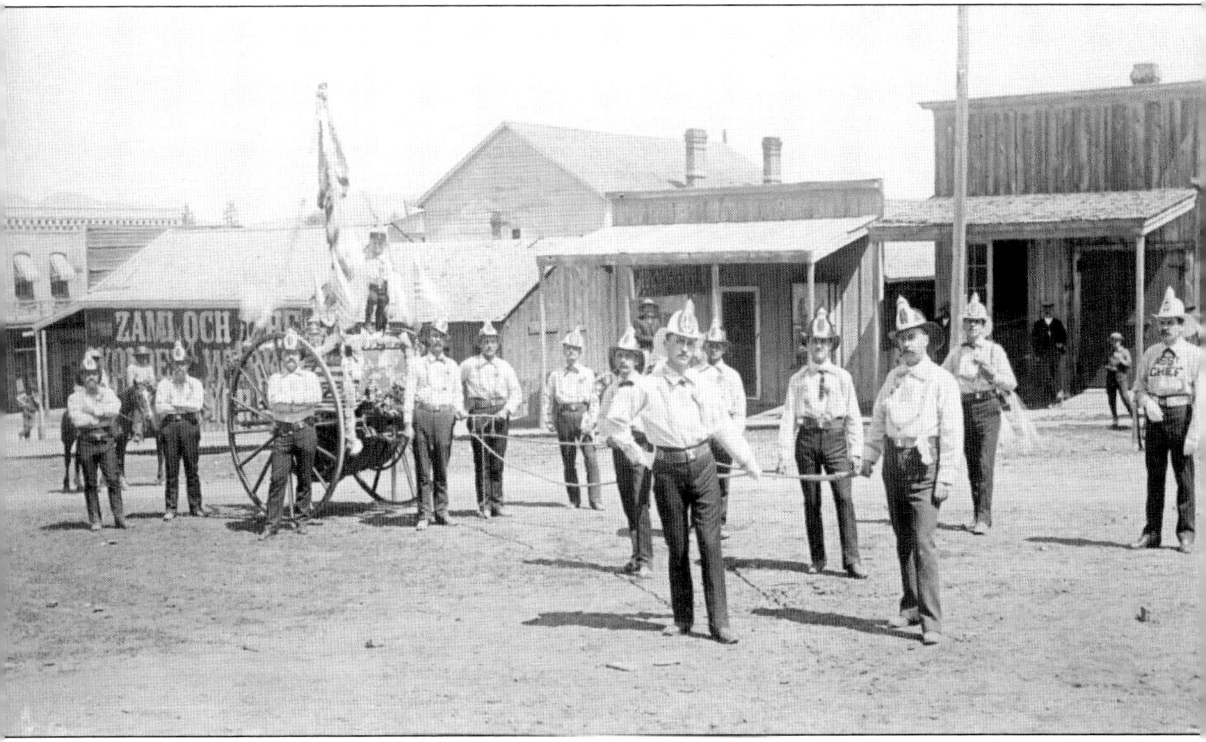

The Toughs No. 1 officially formed on April 2, 1885. The first two chiefs, Al Whitney and Charles Spencer, are listed as honorary members on the roster. Pictured on July 4, 1888, the Toughs No. 1 company has their elaborately decorated hose cart ready. Competition between the four companies covered not just job performance, but also extended into the parade floats as well as the annual hose-cart races. Arguments as to who should be given credit of being first to a fire resulted in the establishment of a set of rules in October 1885, section 19 of which read, "The first company at a fire shall attach to the plug and the second company lending its hose to the other shall have the right to use the nozzle. The company first at the hydrant shall have credit on the record for having first turned on stream." (Courtesy of Sharlot Hall Museum, #fi138p.)

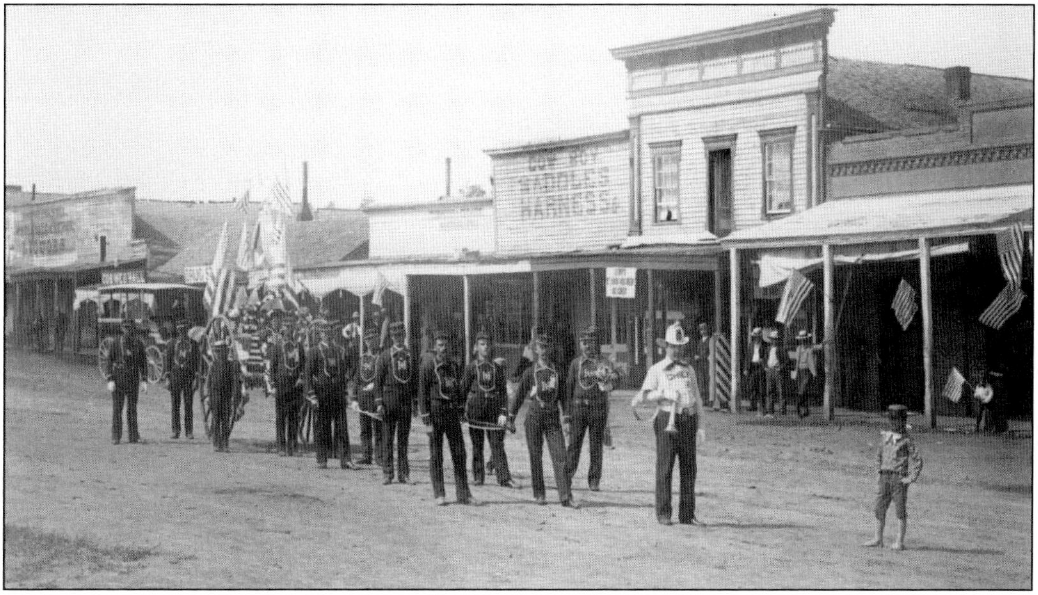

Above, the Toughs No. 1 acquired newer, dark blue uniforms. Standing in the lead of the hose cart is Will Coles Bashford, sporting his new chief lapel. Below, the 1895 team prepares for another Fourth of July contest. The Toughs' brotherly animosity towards the Dudes would start with their very first hose-cart race on May 1, 1885, when the Dudes handily beat the Toughs. Newspaper reports the following day stated that the Toughs had a much-disgruntled supporter who reportedly shattered the expensive French glass window of jeweler George H. Curry, a member of the Dudes. The unruly supporter was a pug dog of "orthodox ugliness" owned by one of the Toughs members. (Both, courtesy of Sharlot Hall Museum: above, #fi110p; below, #fi132p.)

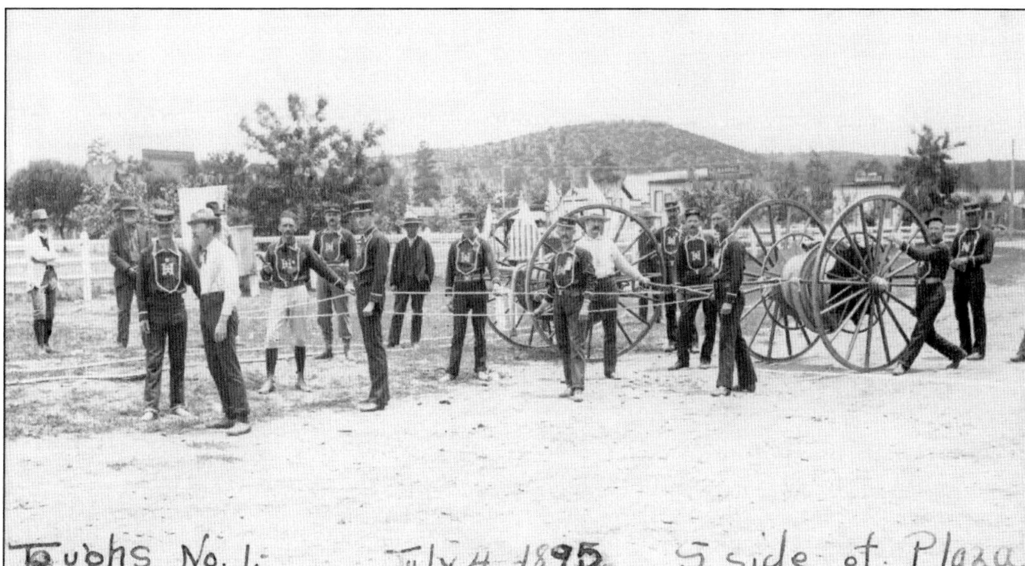

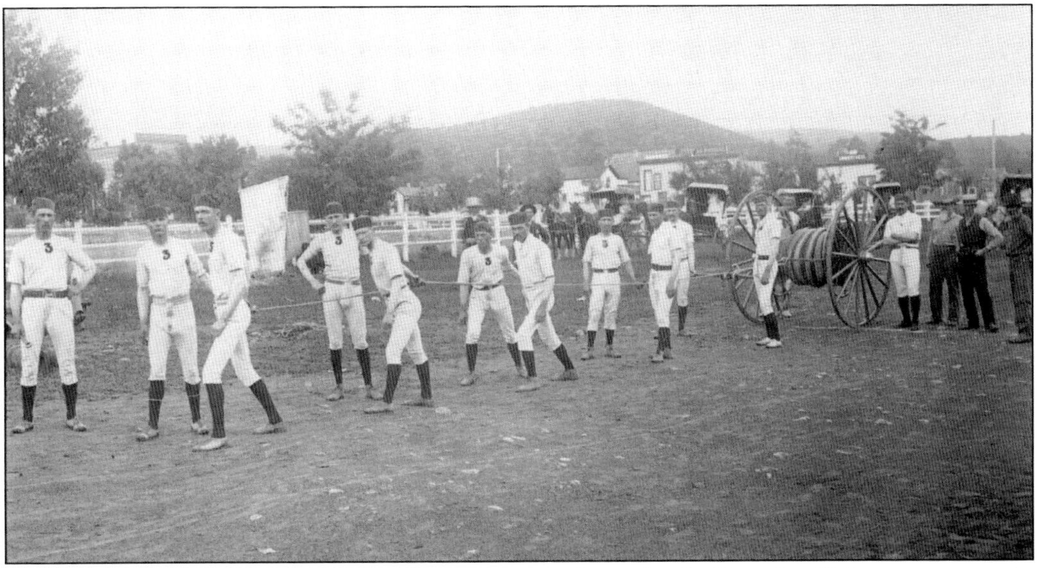

The O.K. No. 3 team is seen here in their racing uniforms for the Fourth of July competition in 1891. Like the Hooks, the team would have had to borrow a cart from one of the other teams. The O.K. No. 3 roster would produce chiefs Robert Connell and Bert Tilton. Martin Testori of the O.K.s would have some notoriety as being the longest-serving firefighter in Arizona in 1921. He came to Prescott in 1891, having already served as a firefighter in his native Switzerland. (Courtesy of Sharlot Hall Museum, #fi130p)

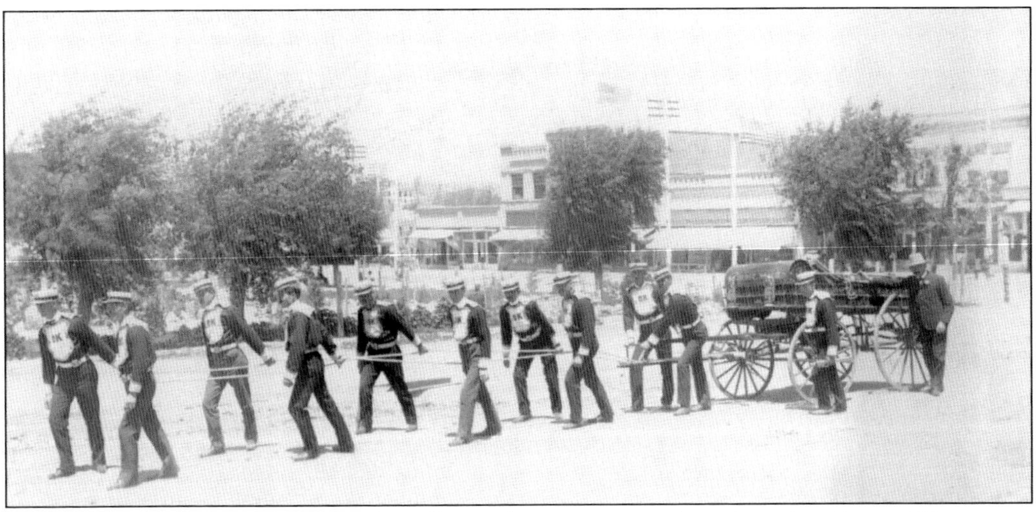

In this c. 1903 photograph, the O.K.s pull their hose wagon in their formal uniforms. The wagon was stored at the northeast corner of Gurley and Fluery Streets. Striking is the fact that the O.K.s had a heavier wagon to contend with than a hose cart, yet they did not make use of the horses stabled in the corrals at the O.K. Stables next door to pull the load. (Courtesy of Sharlot Hall Museum, #fi109p.)

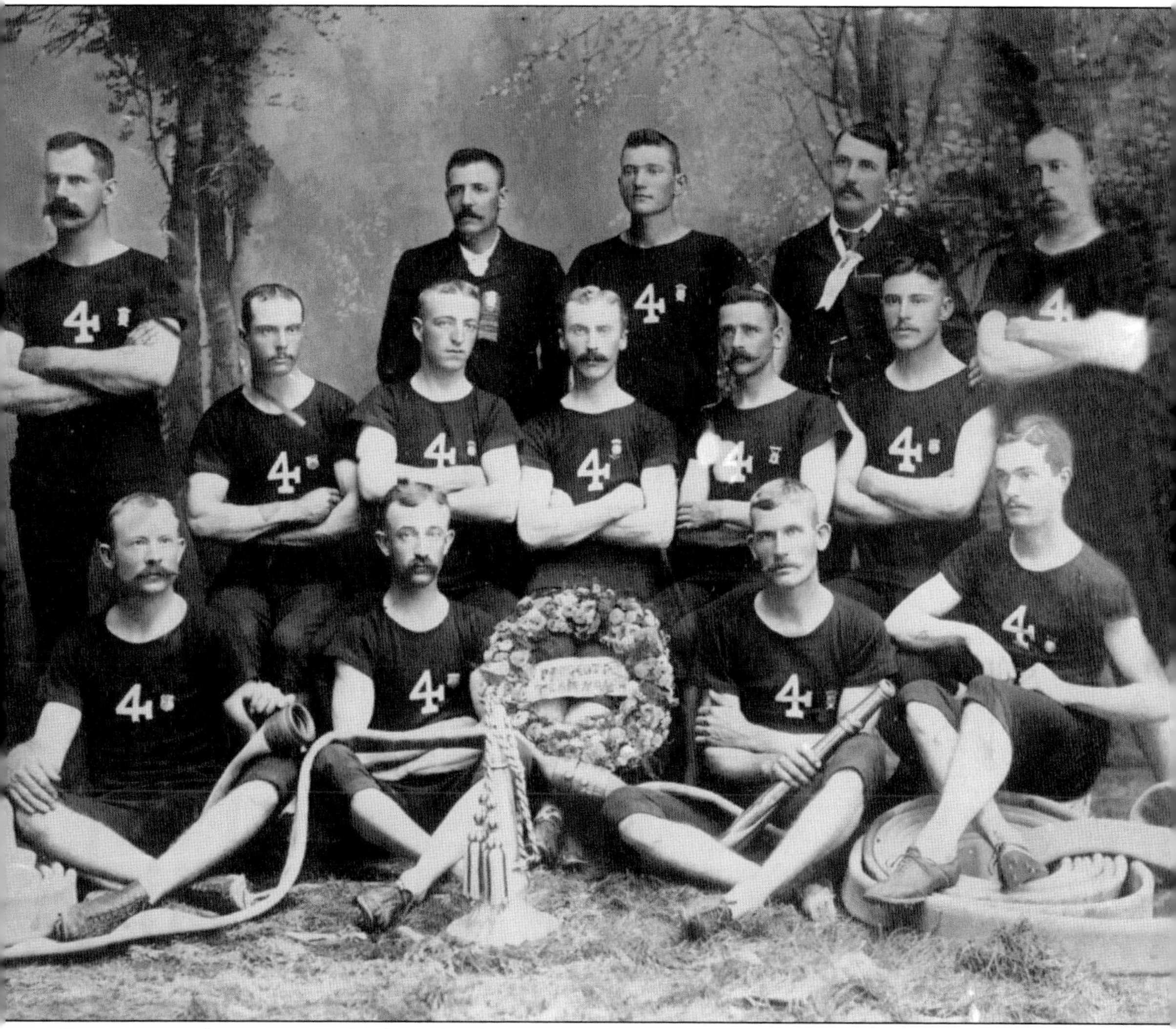

Company rivalry was set aside when the department faced outside competition. Prescott Volunteer Fire Department sent a team to compete in the Southwest Regional Hose-Cart Races held in Albuquerque, New Mexico, for multiple years. The team was comprised of members from all four of Prescott's crews. They took the regional championship three years running, from 1890 to 1892, to the great dismay of the Albuquerque department, which demanded the return of the silver-wrought traveling championship belt, writing letters to the city and department declaring that Prescott's wins were without merit and that the belt should be returned. Will Bashford wrote in response to Albuquerque's complaint, "The belt, having been now won three times in succession will remain in this department forever." In a further act of defiance to Albuquerque's demands, Prescott firefighters sent them a belt of sorts . . . the collar off of Sport, one of the department's dogs. (Courtesy of Sharlot Hall Museum, #fi113p.)

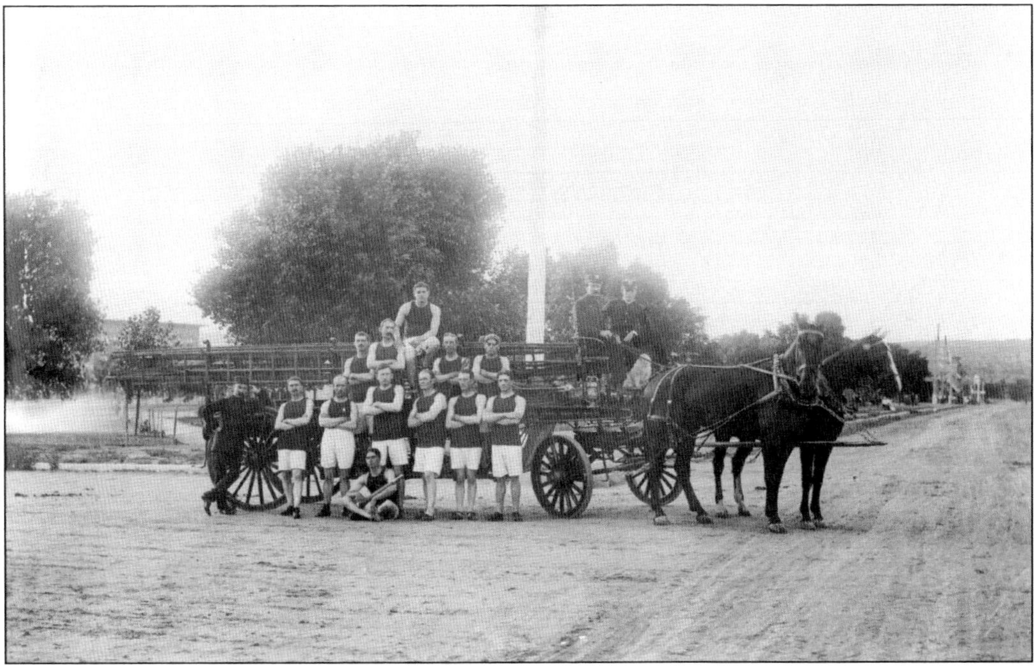

In 1906, the City of Prescott purchased the Howey Hall building on the southeast corner of Goodwin and Cortez Streets for the use of city business and the adjacent building for the stabling of two horses and housing of a hose wagon. The four fire companies were consolidated to the one firehouse and operated out of the building off of the hose wagon, with two hose carts held in reserve there as well. The horses were purchased in Phoenix for $350. The first wagon proved too heavy for the horses on Prescott's steep hills and a second, lighter hose wagon was purchased. (Both, courtesy of Sharlot Hall Museum: above, #135p; below, #fi144p.)

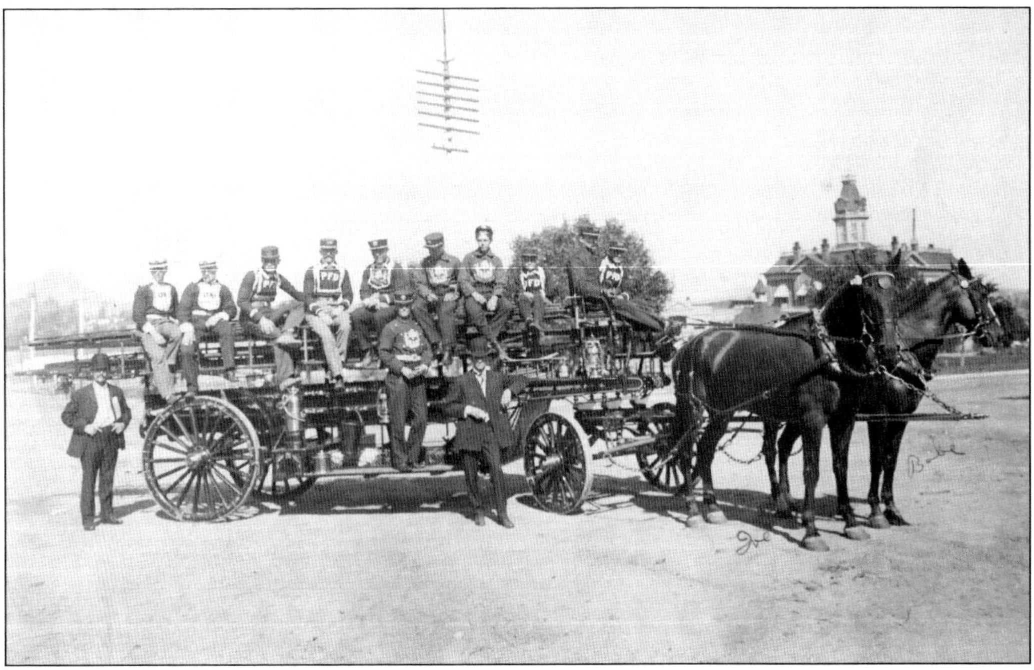

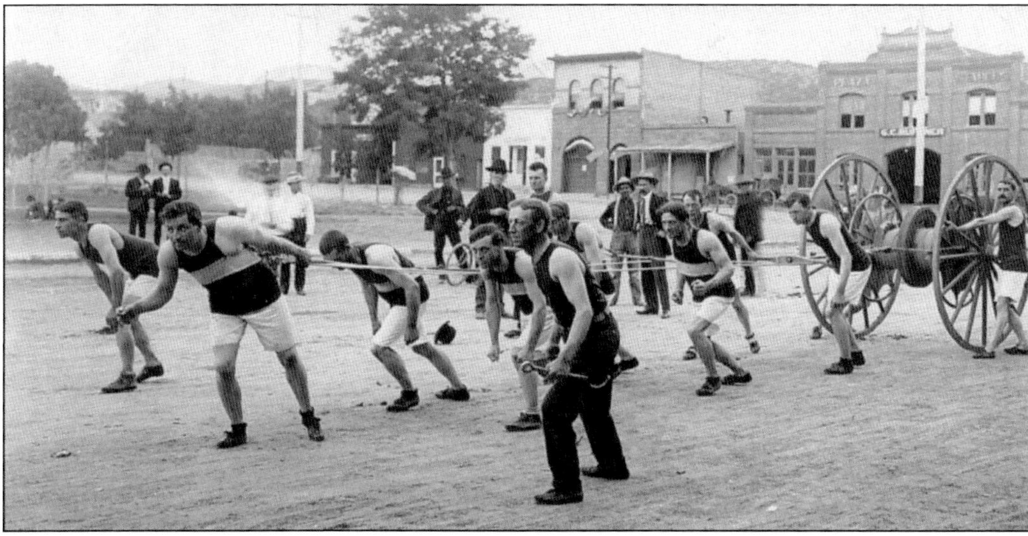

Noteworthy in the background of this 1907 photograph is the current chamber of commerce building (center, with three small arched windows in the second story). The building is often confused with being the first consolidated firehouse for the department. While the building was used as the city jail, it was not a firehouse. The two buildings had similar architectural features, lending to the confusion. (Courtesy of Sharlot Hall Museum, #fi136p.)

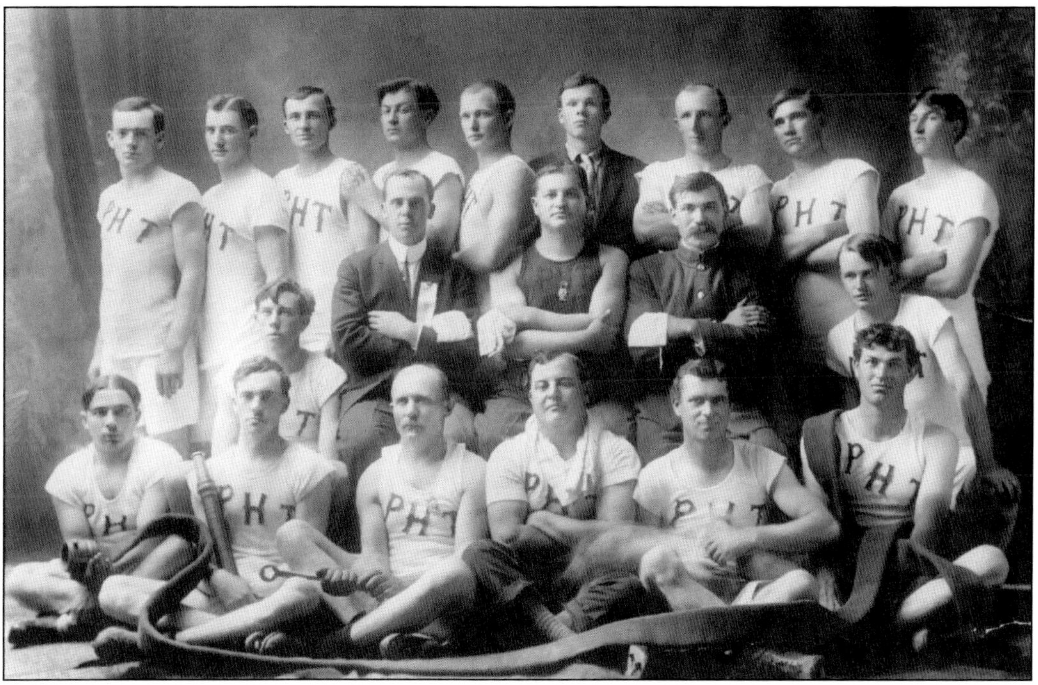

This 1912 photograph shows a combined Prescott hose team formed for competitions. Only Assistant Chief Peter and longtime firefighter Martin Testori are sporting iconic firefighter mustaches. This same team beat their Jerome Fire Department rivals in a Labor Day event, nearly breaking the posted record, a feat for which Chief Tony Johns rewarded them by paying for a banquet for the whole of the department. (Courtesy of Sharlot Hall Museum, #fi119p.)

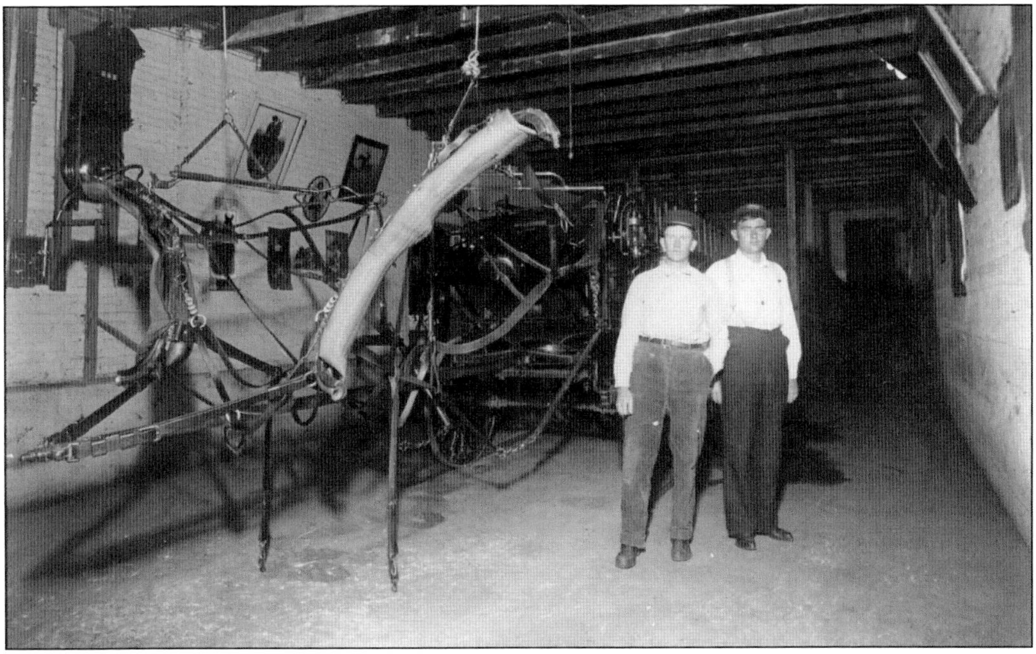

These rare photographs are of the inside of the consolidated firehouse on Cortez Street. The above picture shows how the harness was held ready for the horses. To the right rear in the shadows, one of the horses can be seen. The second photograph shows the horses hitched up and ready to pull. On the wall over the horses' heads is the Gamewell indicator box, which would show which alarm box was pulled. Firefighters had to know the location of every alarm box in the city. This same indicator box currently hangs in Station 71. (Both, courtesy of Sharlot Hall Museum: above, #97.109-13; below, #97.109-12.)

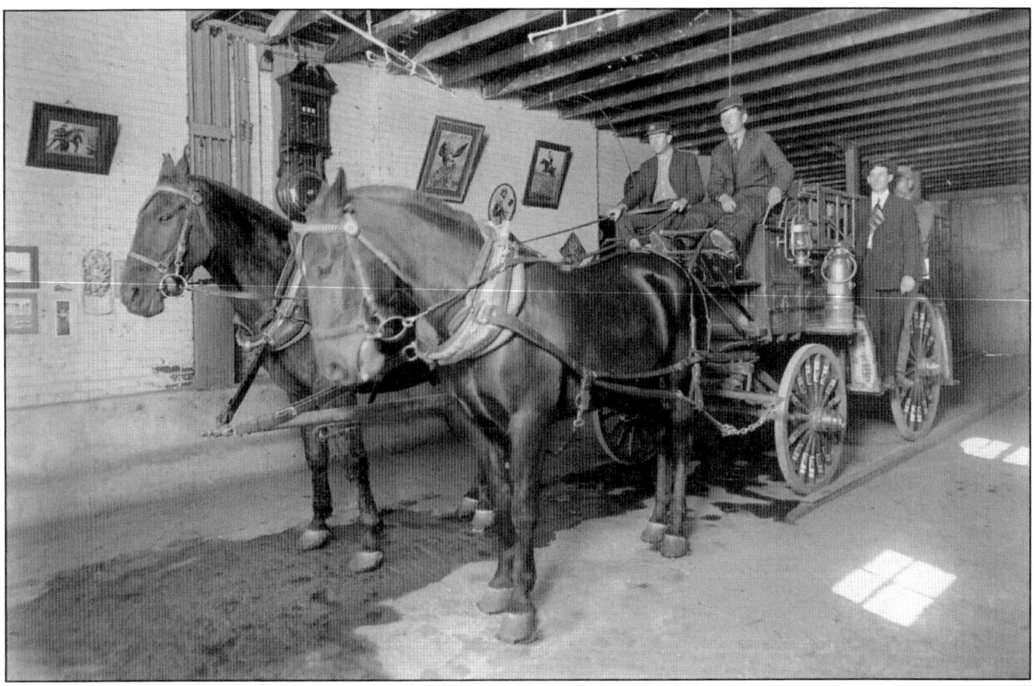

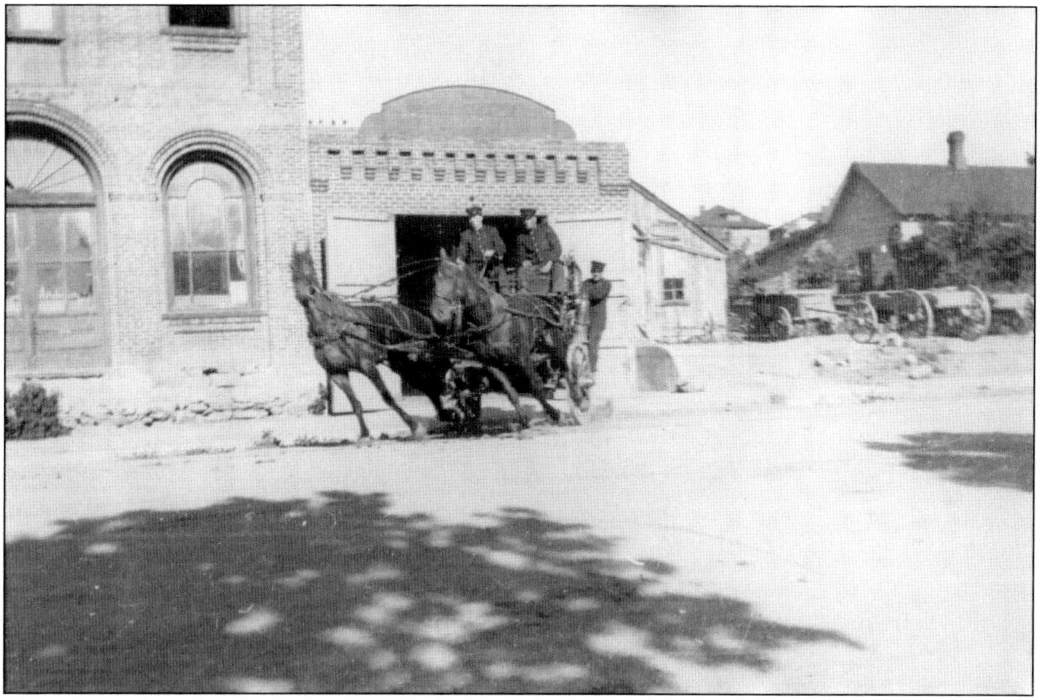

This is one of the few photographs of the horses pulling the wagon out of the station. At the reigns is Jim Fitzgerald. Fitzgerald was one of the first two men on the department who were paid, in charge of caring for the horses and being ready to hitch them to the wagon. Years later, Fitzgerald's nephew Gerald Fitzgerald would be one of the engineers of the motorized fleet. (Courtesy of Sharlot Hall Museum, #97.109-20.)

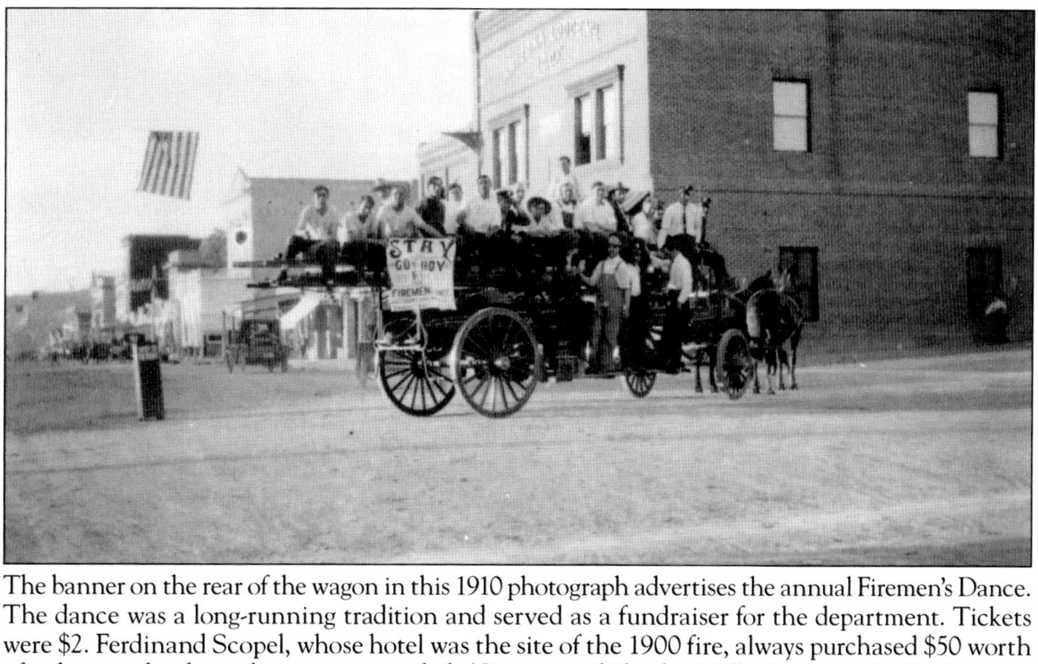

The banner on the rear of the wagon in this 1910 photograph advertises the annual Firemen's Dance. The dance was a long-running tradition and served as a fundraiser for the department. Tickets were $2. Ferdinand Scopel, whose hotel was the site of the 1900 fire, always purchased $50 worth of tickets to the dance but never attended. (Courtesy of Sharlot Hall Museum, #97.109-21.)

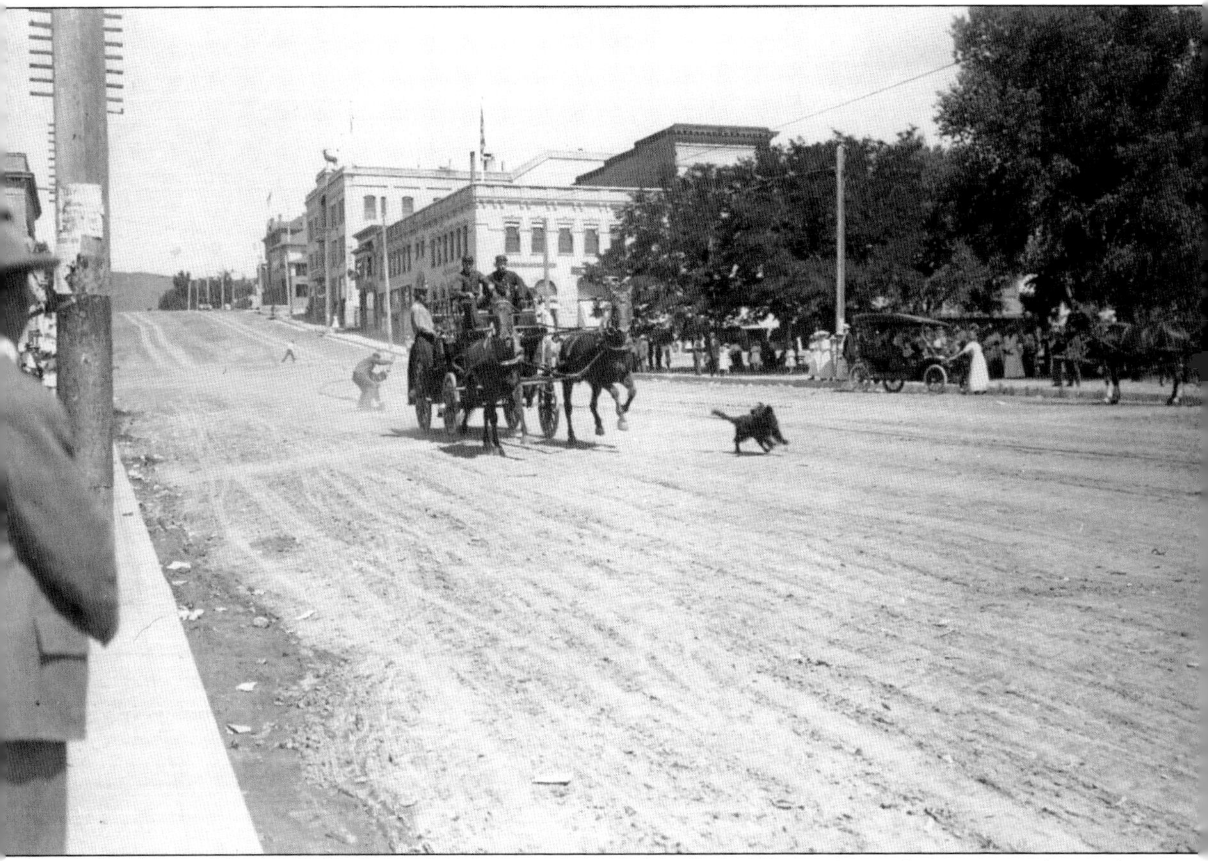

This iconic photograph of the newly purchased and smaller horse-drawn hose wagon shows the firefighters running a practice drill on Gurley Street. The purchase of the horse-drawn hose wagon forced the City of Prescott to hire two engineers who were responsible for the care and upkeep of the two horses. Ultimately, the horses would eat themselves out of a job, as they were consuming over $500 per year in food in addition to the expense of paying the two men seeing to their upkeep. Years later, when the department made the transition to motorized vehicles, the two horses were retired and used to pull the city's trash wagon. They were reported to have run amok with the trash wagon anytime they heard the sounding of the fire bell. (Courtesy of Sharlot Hall Museum, #fi117p.)

Two

A Mechanized Fleet

With the rising costs of feeding the horses becoming an issue, the department embraced the idea of using a mechanized fleet. The first acquisitions were two 1913 Ford Model Ts in 1915. While not truly fire apparatuses, they provided transportation for men and equipment. A 1921 Velie was the next purchase. It too was not a pumper, but rather was more like a mobile fire extinguisher that was intended for incipient fires.

Prescott Fire Department is very fortunate to still have its first two actual pumping fire apparatuses that were purchased by the city housed locally. The first engine pumper purchased was a 1926 American LaFrance. The first ladder truck was a 1931 Seagrave. The American LaFrance is still in the possession of the Prescott Fire Department. The Seagrave was sold to the local Prescott Antique Auto Club. Both apparatuses still operate and function primarily as parade vehicles.

The LaFrance saw service from 1926 all the way through to 1954. Its age at the time of its retirement was used to leverage the purchase of a new engine as the insurance underwriters of the time were starting to take note of the antiquated equipment that was still being touted as a front-line engine. The city council acquiesced and approved the purchase of a 1957 Seagrave triple combination pumper.

The story of the life of the 1931 Seagrave ladder truck follows a decidedly different path. The Seagrave truck saw service on the Prescott Fire Department until 1984, at which point it was sold to the Prescott Antique Auto Club for a dollar. The aged vehicle bounced around to the different yards of the club's members for over 20 years before the club made the definitive move to restore the dilapidated vehicle. The Seagrave was then brought into the club's Quonset hut and the refurbishment began. The vehicle was stripped down to its frame, repaired, repainted, and reassembled again. The process took over three years. The result has been nothing short of breathtaking, and she has been a hit in the parades ever since.

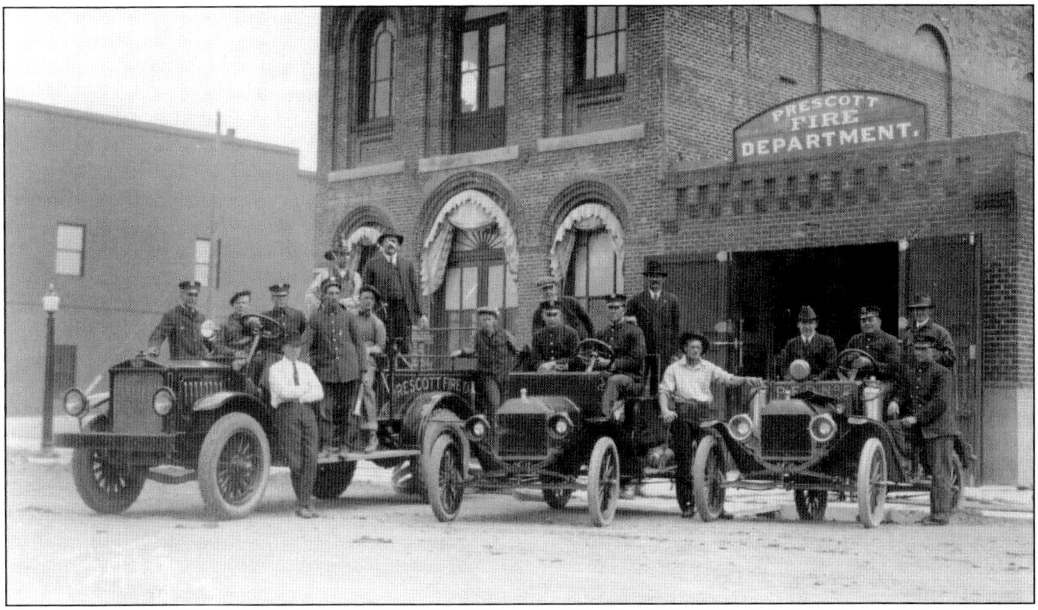

In 1922, with the arrival of the 1921 Velie, the vehicles were assembled and photographed in front of the station. The 1921 Velie was the first apparatus purchased as a true fire apparatus, but it had no pump, only a 70-gallon extinguisher that was plumbed into a hose reel, with the idea being that incipient fires could be doused with an extinguisher instead of copious amounts of water. This would help to reduce the subsequent and inevitable water damage. The Model Ts were identified as "1" and "2," with the Velie being "3." (Both, courtesy of Sharlot Hall Museum: above, #fi153pb; below, #fi153pc.)

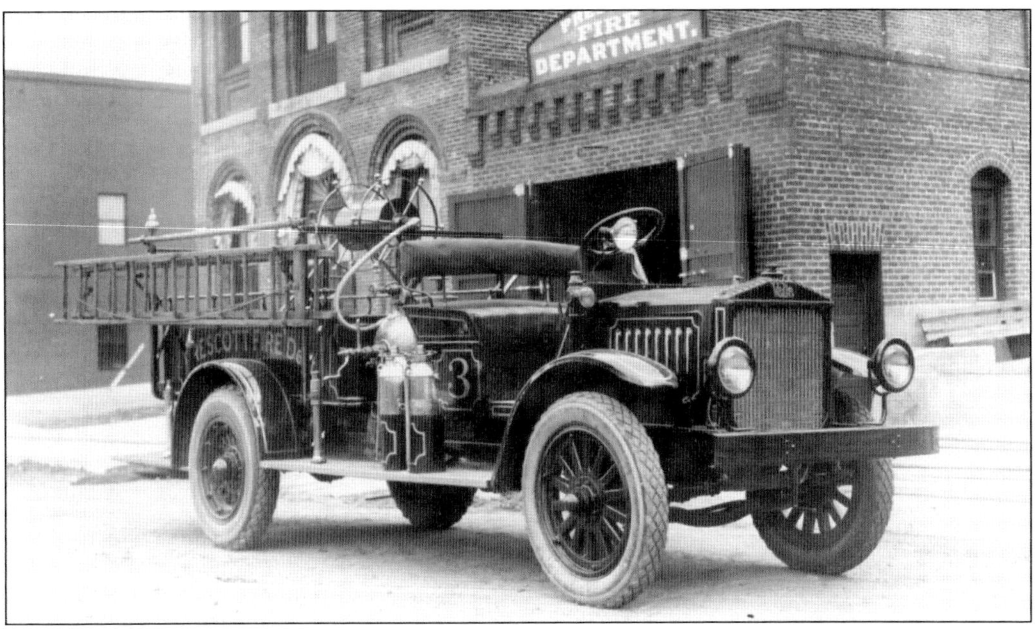

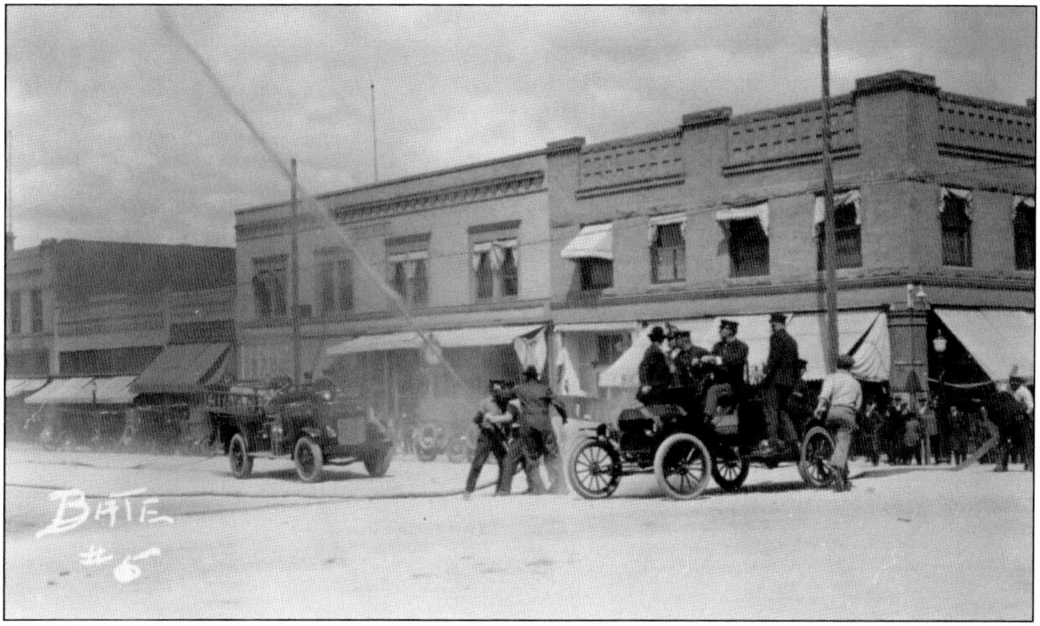

The firefighters continue a decade-old ritual of Sunday practice. Firefighters of the three engine companies—the Dudes, the Toughs, and the O.K.s—would test their ability in seeing how fast they could hook three hydrants and deploy their hose lines as they circled the courthouse square. In 1915, the department won a $100 bet that they could not break four minutes to complete the task. The men astounded the betters by accomplishing the task in three minutes and three seconds. Here, the firefighters work the same route as their predecessors, only using the department's three mechanized vehicles. (Both, courtesy of Sharlot Hall Museum: above, #fi125pa; below, #pb009f5i7.)

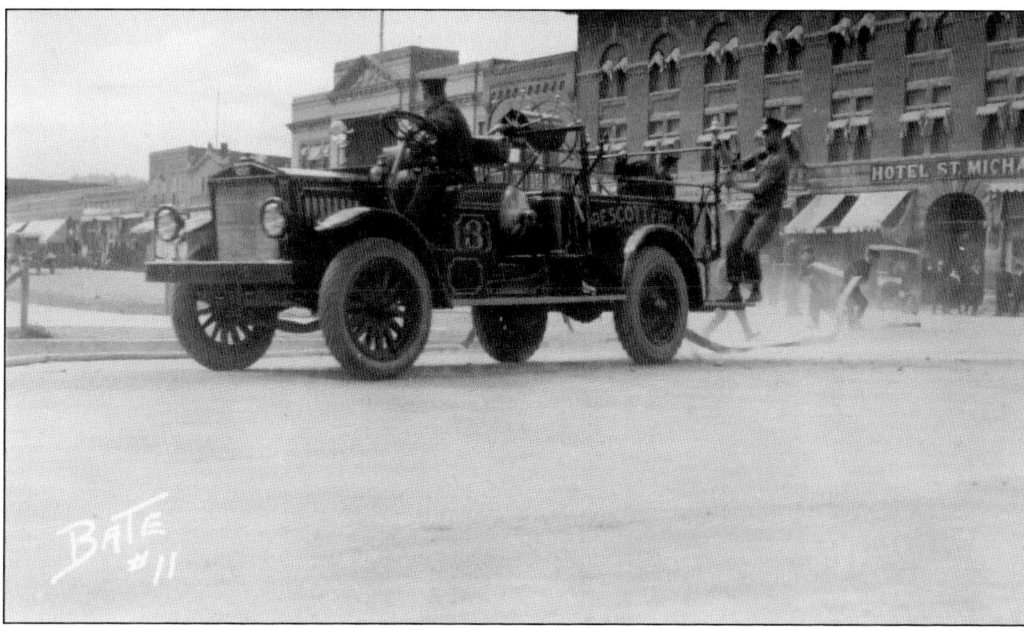

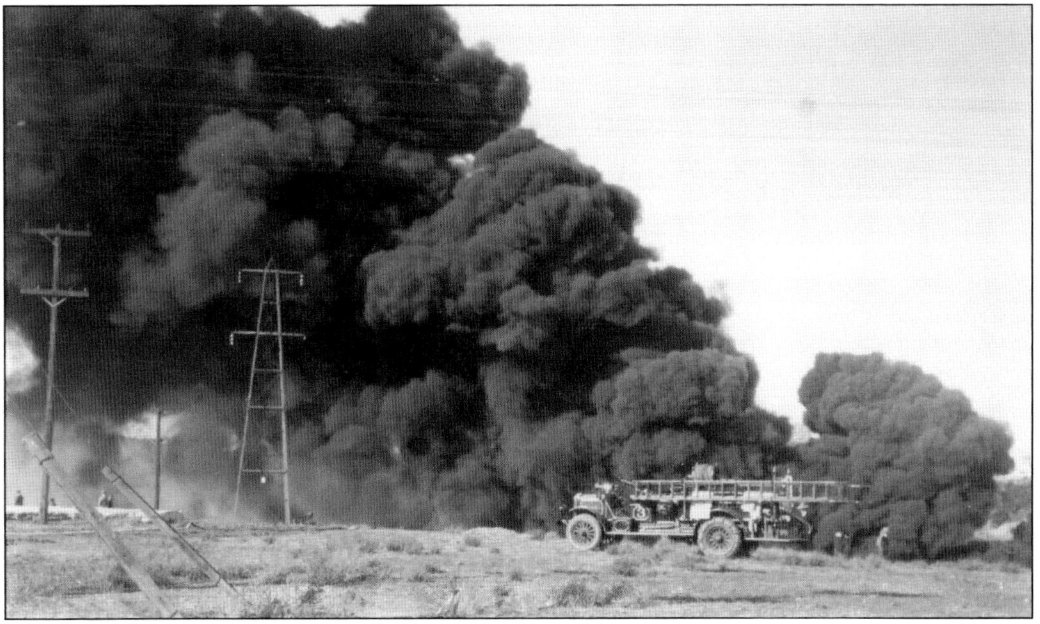

The Velie's extinguisher tank was most effective on incipient fires, not a blaze like the one seen here. The Velie had a hose bed full of two-and-a-half-inch hose, which was run directly off of the hydrant. (Courtesy of Sharlot Hall Museum, #fi129pa.)

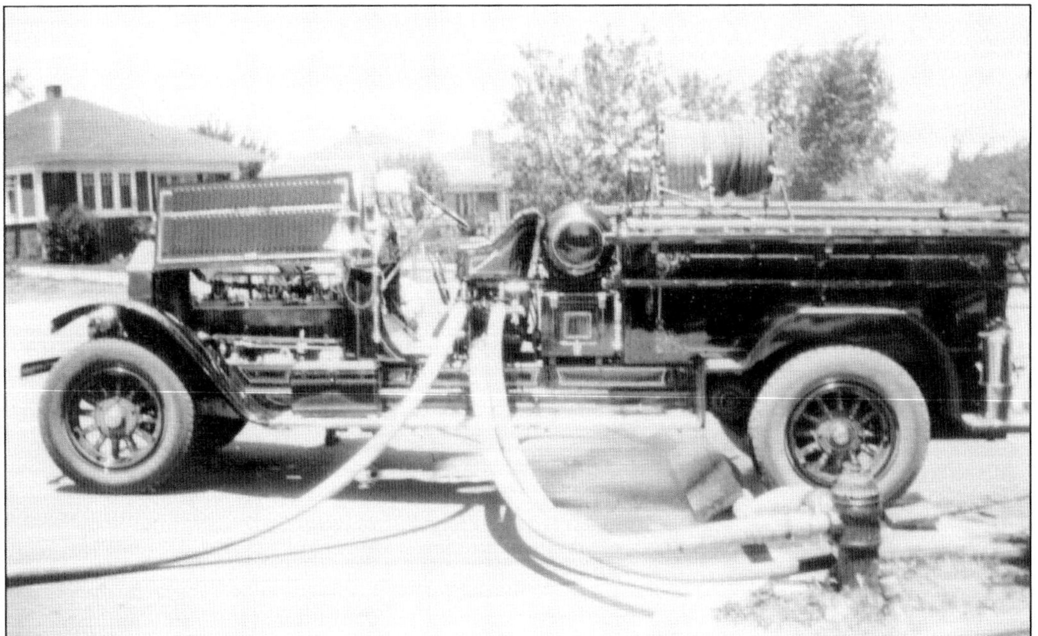

While at first glance it would appear that the newly arrived 1926 American LaFrance is pumping a large fire from a hydrant, this image actually captures the LaFrance undergoing a pump test to verify its 750-gallon-per-minute pumping capacity. Pump capacity testing occurs annually with all the engines in the fleet to this day. (Courtesy of Sharlot Hall Museum, #fi129pf.)

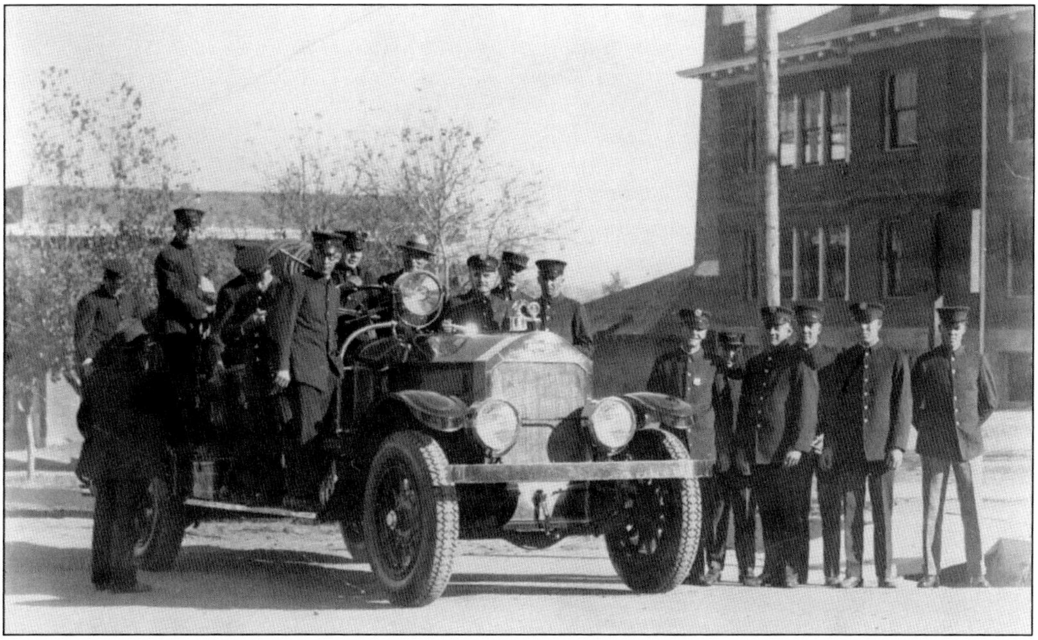

Taken in front of Washington School, this 1927 photograph shows the new American LaFrance 75. Standing on the running board on the left are Morris Payne, Ross Lane, and Bob Connell. At the wheel is Gerald Fitzgerald, a paid engineer. Below, the firefighters demonstrate the pumping ability of the new engine on the corner of Gurley and Montezuma Streets. The LaFrance was quickly nicknamed the "Frog," likely due to the French slur, or possibly because of the way it tends to hop when changing gears. Irony would come some 60 years later when one of the Frog's primary operators was engineer Scott MacDermid, whose last name was quickly distilled down to "Kermit." Yes, Kermit drove the Frog. (Both, courtesy of Sharlot Hall Museum: above, #fi148p; below, #fi101p.)

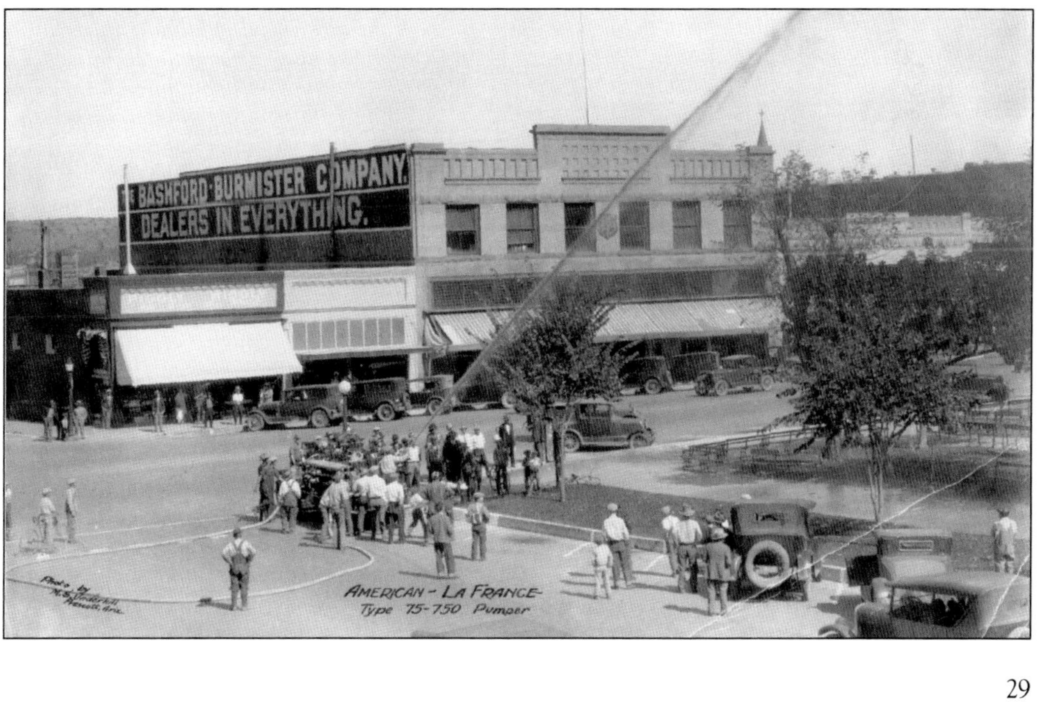

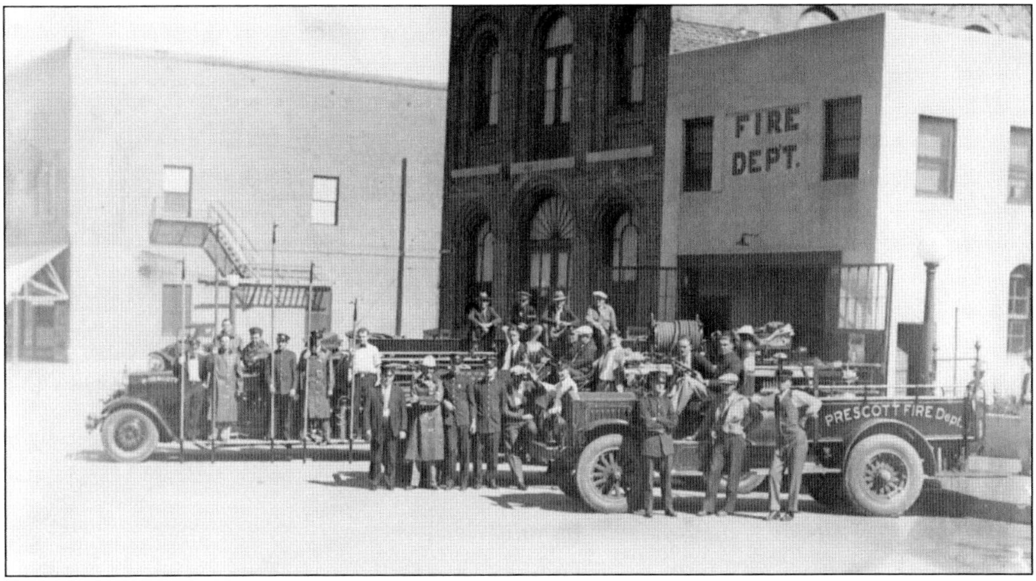

Seen here in front of the Cortez Street station is the fleet with the newly arrived Seagrave truck company to the left, the American LaFrance, and, lastly, the Velie. The two Model Ts were retired with the arrival of the truck company. The Velie remained on the fleet roster into the mid-1940s. Notable too is the addition of a second story to the building, which occurred in late 1929. (Courtesy of Sharlot Hall Museum, #bub8119pc.)

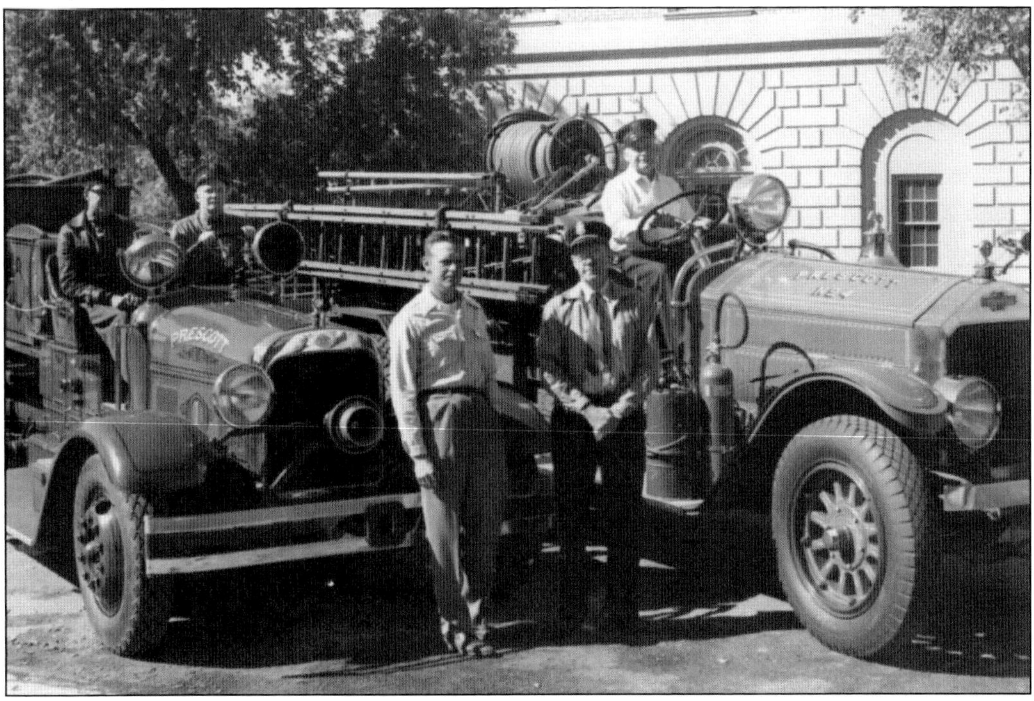

Originally, the American LaFrance was given the fleet number of "4," as seen here, with the two Model Ts being "1" and "2" and the Velie "3." The LaFrance was later changed to "1" after the Velie was retired. The faded "4" can still be seen today. The Seagrave truck was given a "1" as it was the first truck company. (Courtesy of Sharlot Hall Museum, #fi153pd.)

This is the original shipping photograph provided to the department by Seagrave prior to the vehicle's delivery. The truck was sent by train in a boxcar. Joe Mackin, a boy and witness at the time of its delivery, would later have the privilege of driving Ladder No. 1 when he joined the department, replacing Eddie Sweeney as one of the paid drivers. (Courtesy of Prescott Antique Auto Club.)

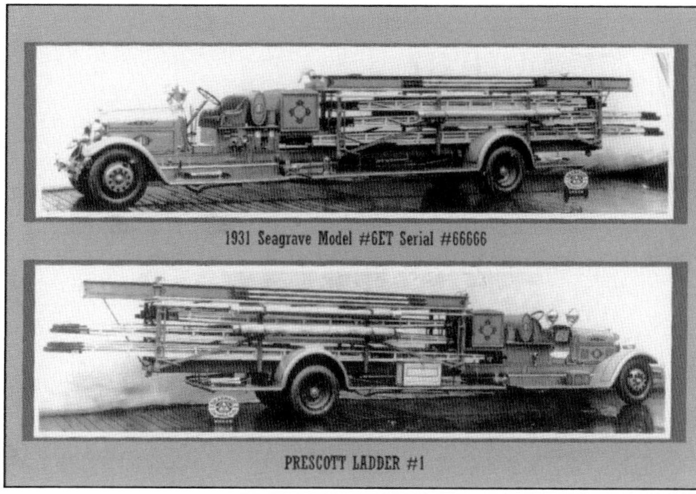

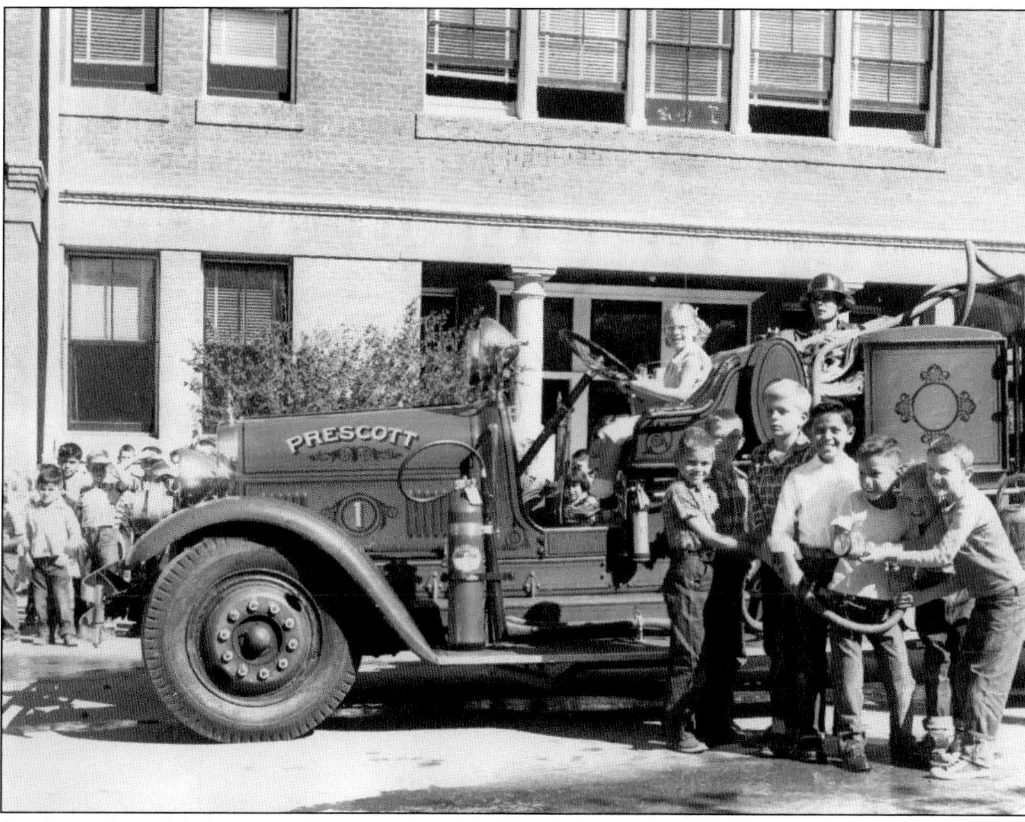

Students from Washington Elementary School are given a chance to spray the hose of the 1931 Seagrave truck. In addition to promotional competitions, Prescott firefighters in the 1950s annually conducted activities at the elementary schools in October as a part of their fire prevention promotions for the month.

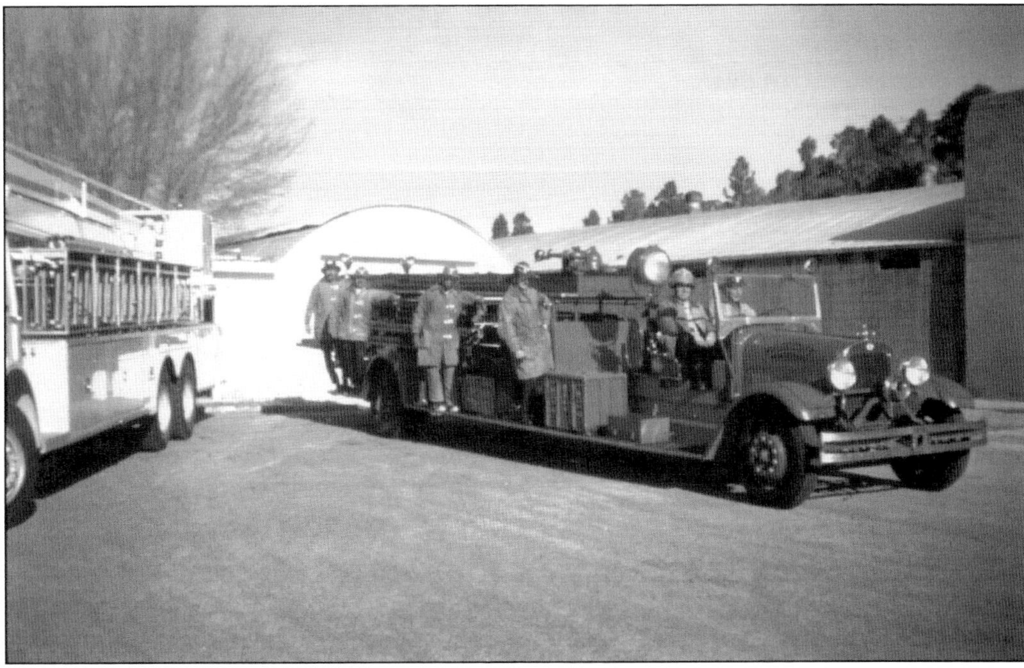

The 1931 Seagrave truck above was eventually dropped to reserve status after the arrival of the Snorkel and moved to Willow Creek Station No. 2, where it would remain until formally retired. Once retired and sold to the Prescott Antique Auto Club, the Seagrave truck remained in several members' yards, the lion's share being in the yard of Rich Eckert, who is seen below driving the Seagrave before the renovation. (Below, courtesy of Janet Travis.)

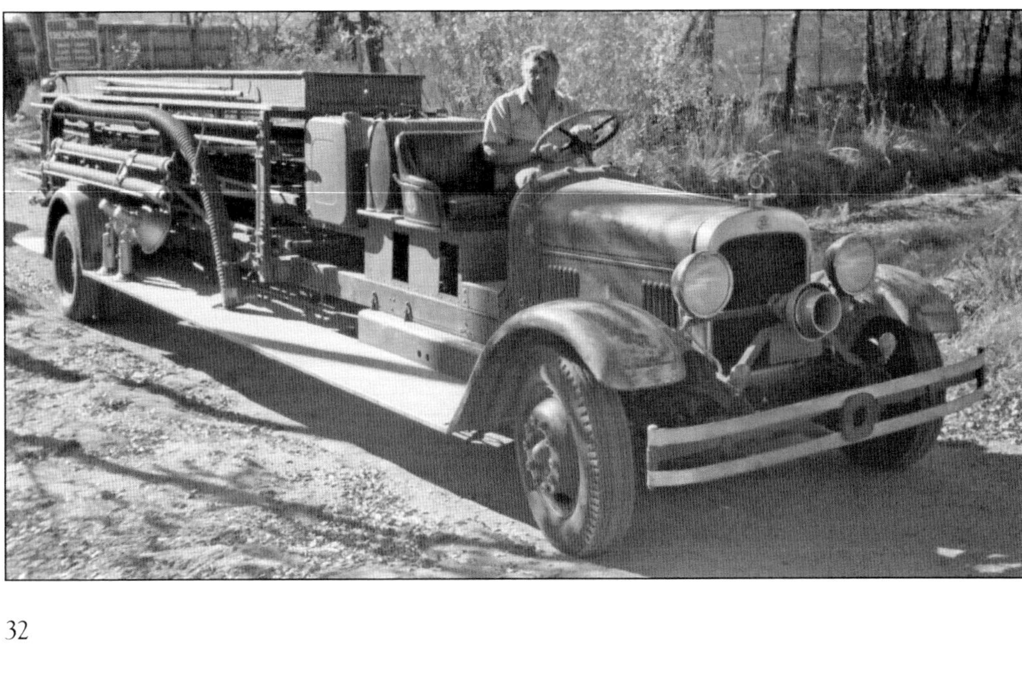

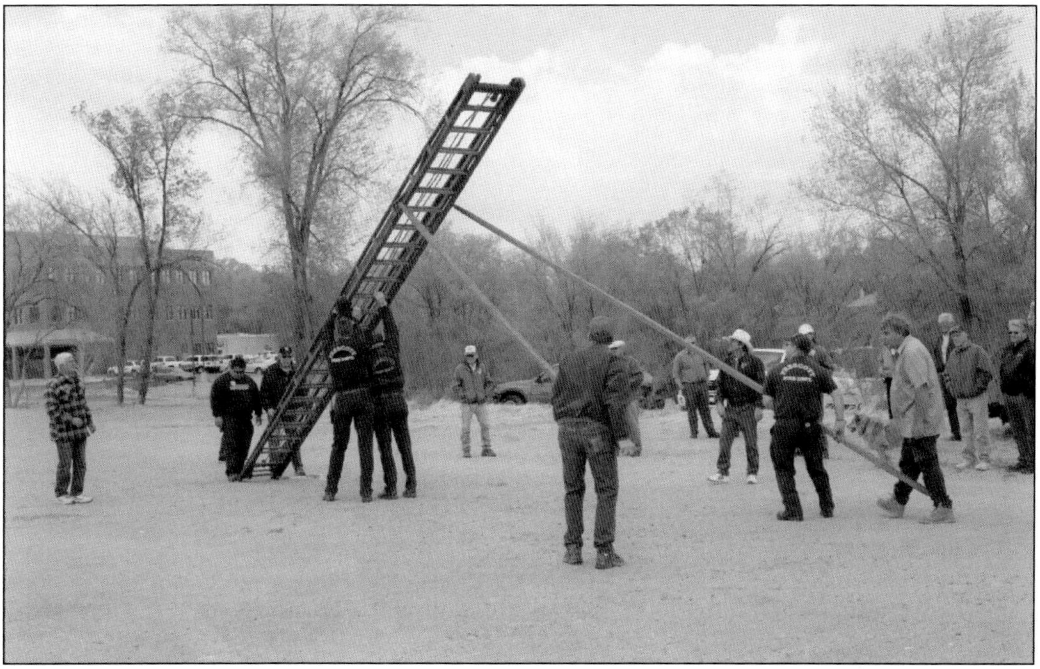

On the Seagrave's moving day into the Quonset hut, retired Prescott firefighters Bob Hinshaw and Ed Hoffman wished to demonstrate the use of the 45-foot wooden extension ladder from the truck, complete with tormentor poles, to the attending current firefighters J.P. Vicente, Bill Hickey, Tom Pleitgen, and Dave Peterson. Hoffman had every intention of scaling the entire ladder in a completely vertical position, held only by the tormentor poles, much as he did during his employment. His wife, Loretta, caught word of his antics and ordered him off the device. (Both, courtesy of Janet Travis.)

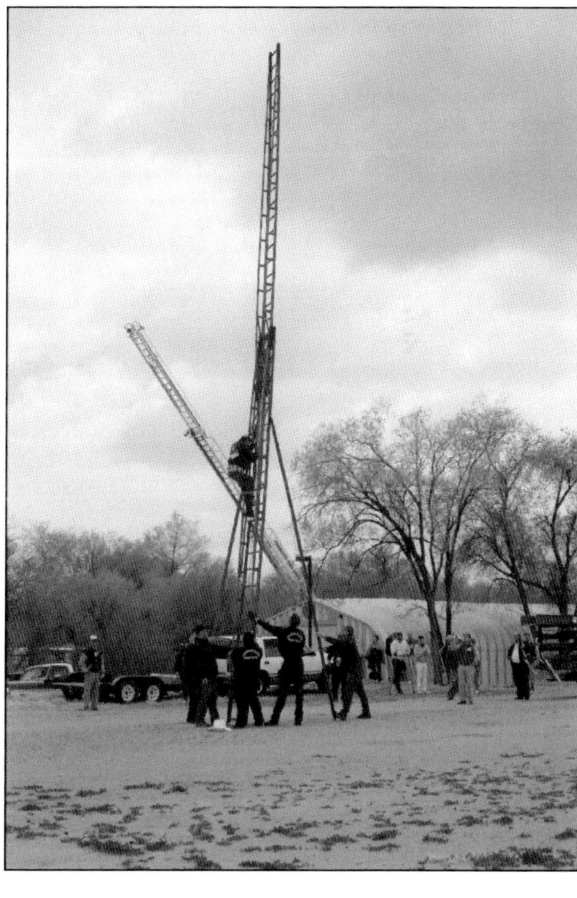

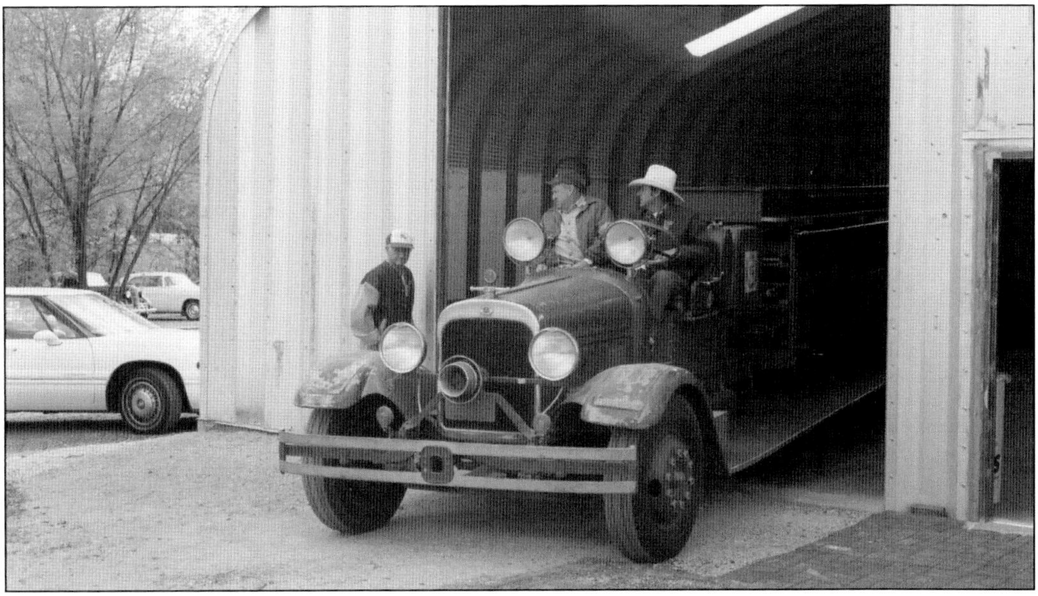

Above, Ed Hoffman, with Jim Kelso joining him on the seat, parks the Seagrave truck in the auto club's clubhouse behind the rodeo grounds. The Prescott Antique Auto Club stripped the Seagrave down to its frame before sanding and repainting every component on the vehicle. Of its original 12 ladders, all but 2 have been recovered, refurbished, and returned to the vehicle. At one point, the original six-cylinder Hercules motor had been removed and replaced with a Buick V-8. The club tracked down a replacement Hercules in order to return the vehicle to its original specifications. Other later modifications, such as a windshield, were removed. The club is still hunting for the vehicle's original bell. (Both, courtesy of Prescott Antique Auto Club.)

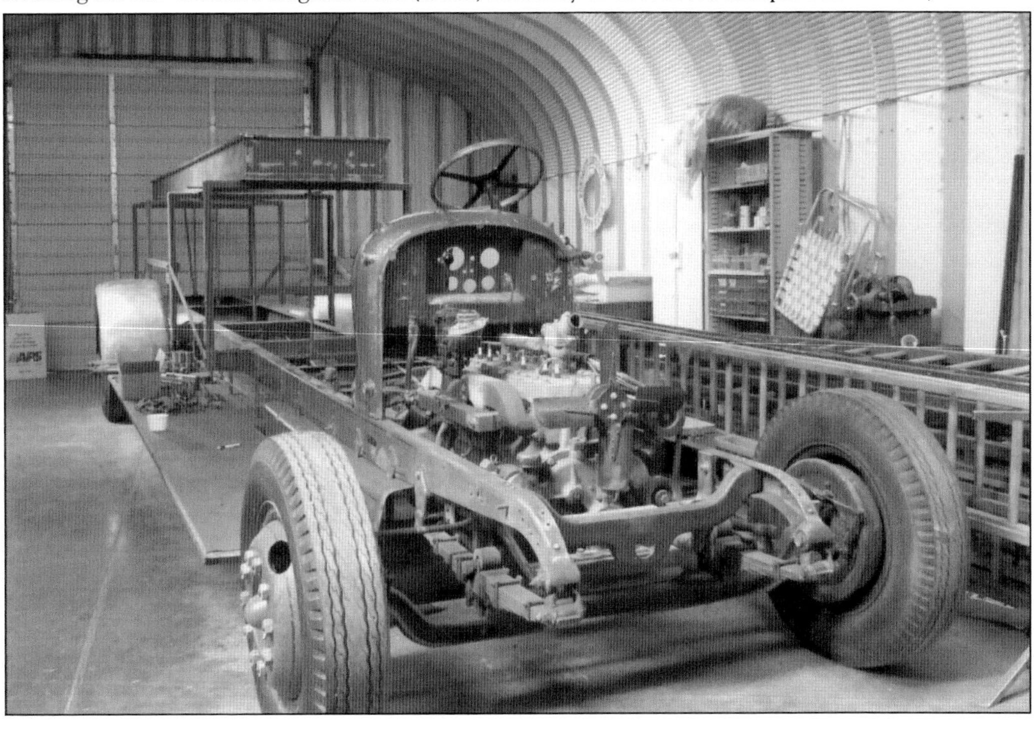

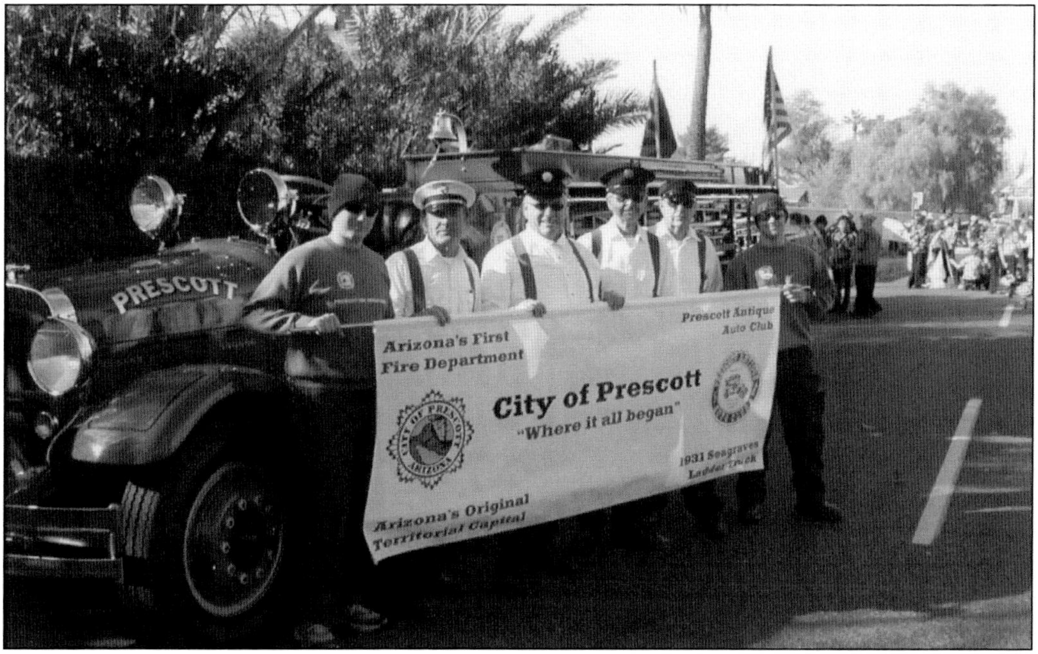

Since its restoration, the Seagrave truck has made multiple appearances in parades. Above, members of the Prescott Fire Explorers Andrew Payne (far left) and Peter Hriljac (far right), along with members of the Prescott Antique Auto Club (from left to right in white shirts) Ed Hoffman, Jim Schultz, Bob Hinshaw, and Charles Rulofson prepare for the 2012 Fiesta Bowl Parade in Phoenix. The trip required a low boy furnished by American Towing to transport it to Phoenix. Below, Prescott Antique Auto Club members Charles Rulofson and Bob Hinshaw enjoy the fruits of their labor, driving the truck in the Frontier Rodeo Days Parade in 2013. (Below, courtesy of Les Stukenberg.)

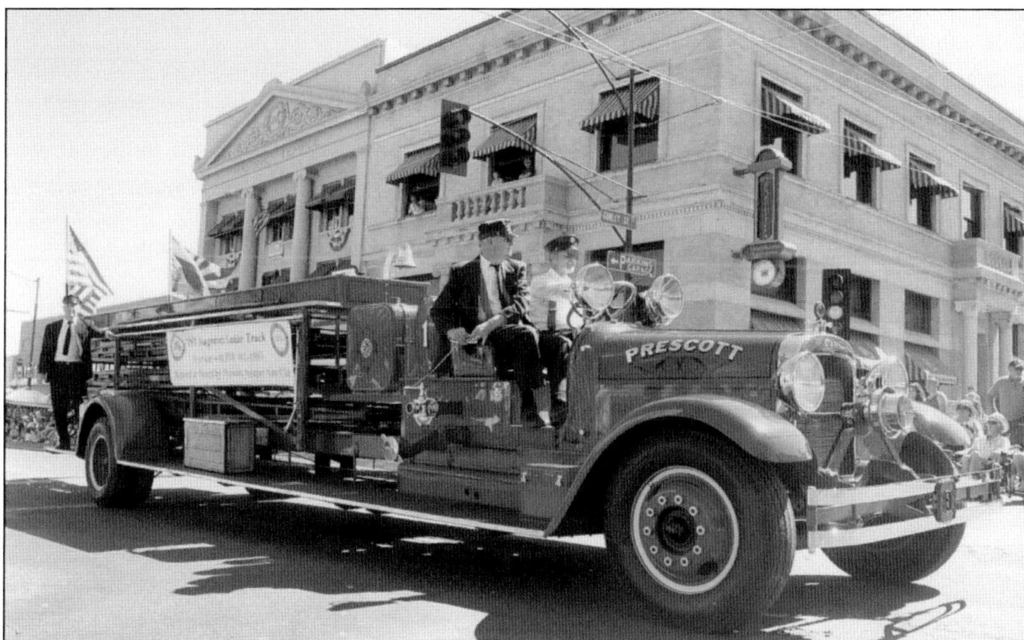

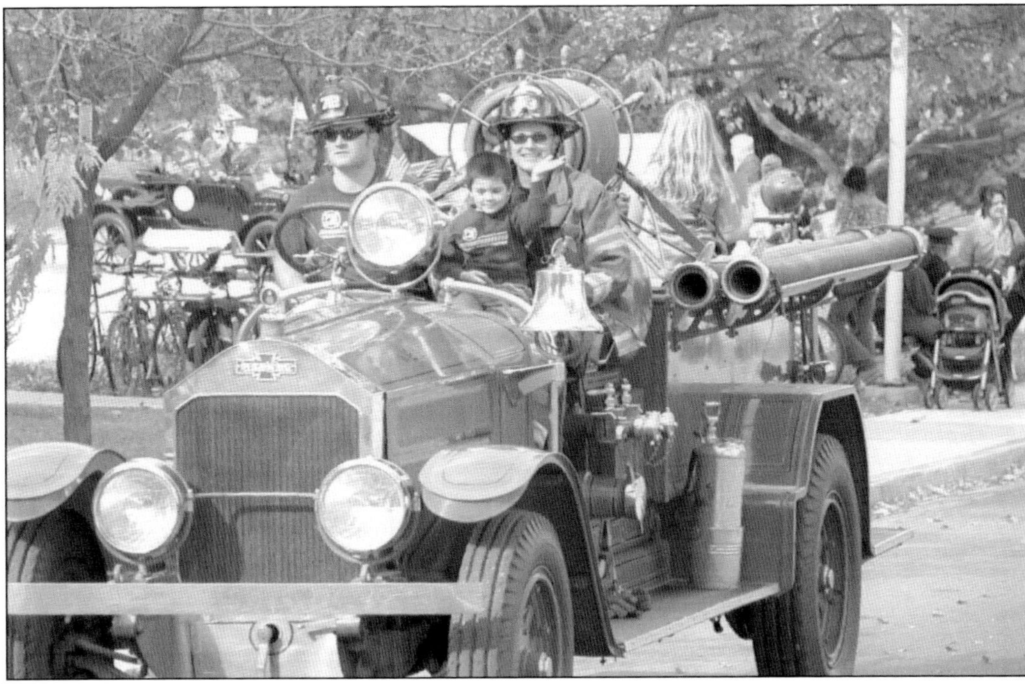

The Frog also enjoys a steady stream of parade rollouts as well as other special events. Above, the Frog is seen driving in the 2006 Veteran's Day parade at the Bob Stump Veterans Administration Hospital. Driving is Lee Montgomery, with Matt Fifield seated next to him. Fifield's son Neil sits in his lap, waving to the crowd. Below, Tera McElwee, daughter of retired battalion chief Tim McElwee, poses with the Frog prior to her and her bridesmaids riding the engine to her wedding ceremony.

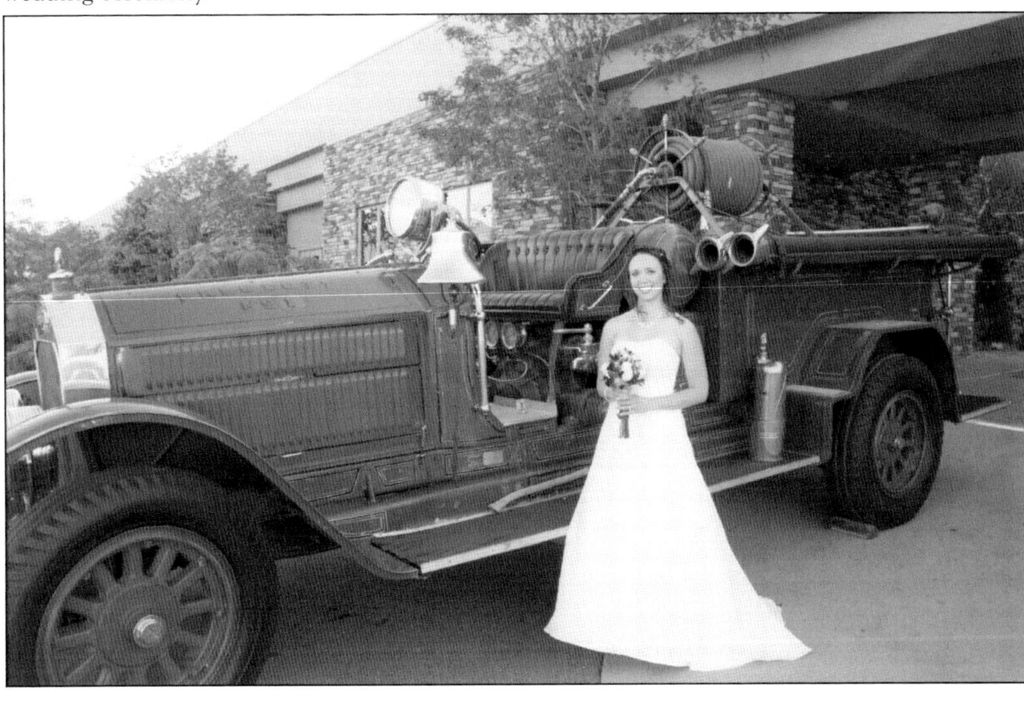

Three

The Changing Stations and Fleet

The Prescott Fire Department started with simple hose carts pulled by the men. Hose carts would be the main mode of response for over 20 years, responding from their various barracks near the courthouse and the center of the city. In 1906, the city purchased a horse-drawn wagon, which was housed in the building next door to Howey Hall in what was formerly the Howey Grain and Feed building. The two buildings had been acquired by the city and were being used for city offices. The Howey Hall building served its purpose for 50 some years.

A growing city with increased fire and police needs created the impetus for a larger facility. A public safety station sprang from that need and was built at Goodwin and Granite Streets. Over time, annexation to all four cardinal points would eventually create the need for additional stations, along with engines and men. Today, the city of Prescott is protected by five Prescott Fire Department stations and one Central Yavapai Fire District station.

With the new stations came new equipment. The first acquisitions were tentative, with the city acquiring war surplus vehicles to round out the fleet. Later, the department would generate specifications to better meet the needs of the city and the firefighters. With almost a century having passed since the first motorized vehicle entered the fleet, the advancing technology abounds. Fire engines and trucks today have mobile computer systems and vehicle locating devices, which allow for faster responses, better exchanges of information, and advanced mapping programs, all of which improve firefighters' responses to citizens' homes. Counting the Frog as the first engine and the 1931 Seagrave as a truck, Prescott Fire has seen the use of a total of 17 engines, 5 trucks, and 3 aircraft rescue vehicles in the past 100 years. The original fleet of three vehicles grew to a current fleet of 16, counting the various special operations vehicles.

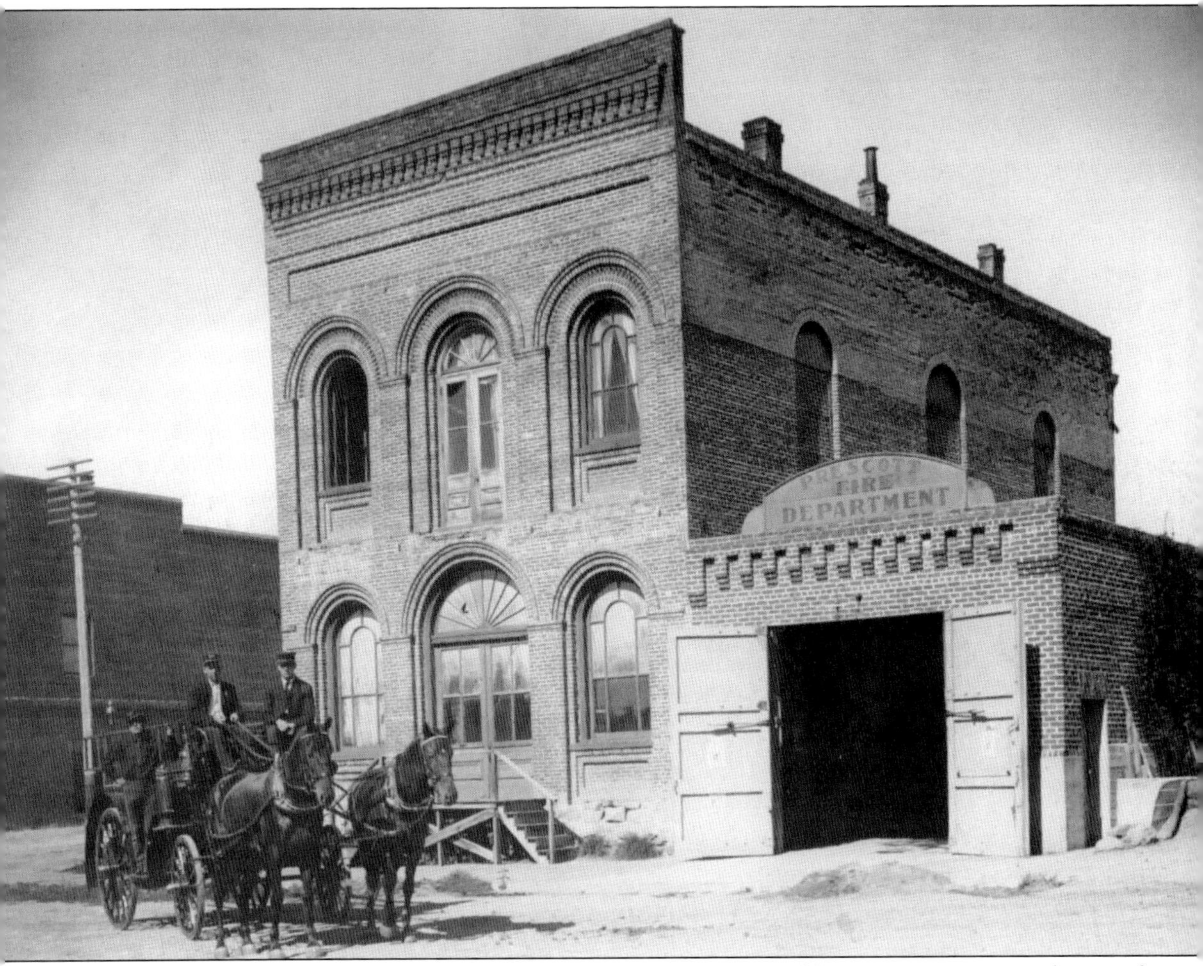

Howey Hall was built in 1876 for blacksmith James Howey. Goldwater ran a mercantile out of the building before moving to a new location. The hall was later purchased in auction by Levi Bashford in 1882. The building would eventually sport a small theater on the second story and an ice skating rink on the first floor. The city purchased the building in 1904, and in 1906, the fire department consolidated its four companies into the building and the adjacent smaller stable building. Insurance maps from 1910 show the buildings housing two hose carts, two hose wagons, and the Hook and Ladder's wagon. At the time, the O.K.s' wagon was still stored at Fluery and Gurley Streets. (Courtesy of Sharlot Hall Museum, #pb124f6i15.)

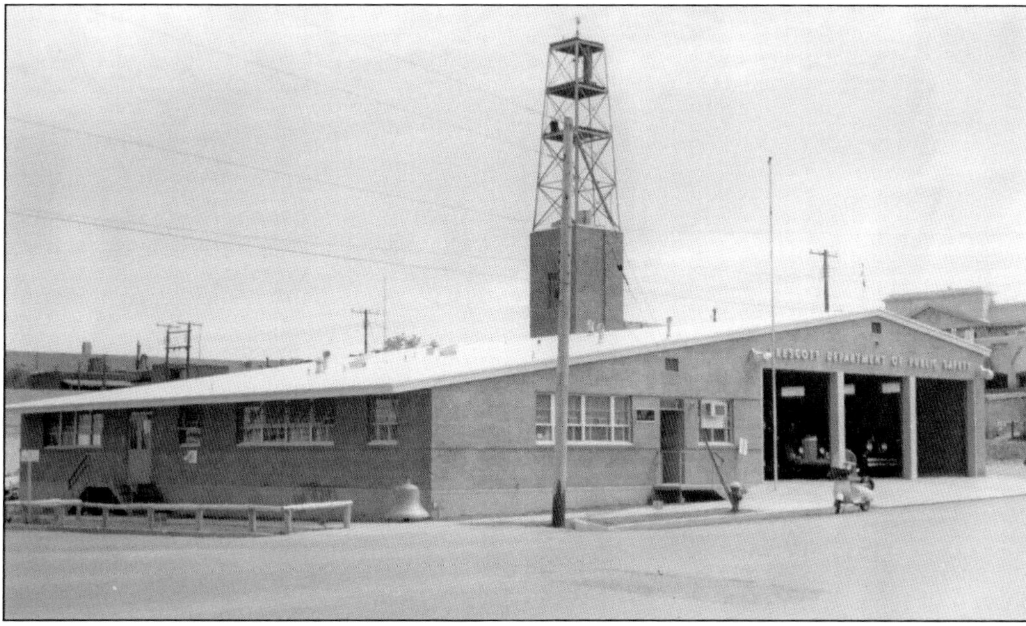

Above is the newly constructed Public Safety Building, located at the northeast corner of Goodwin and Granite Streets. The station was constructed in 1956 to house both fire and police, with the police dispatchers housed in the building as well. They dispatched calls for both departments. Mounted at the corner of the building is the old Maneely bell, which had been previously used to alert firefighters to fires.

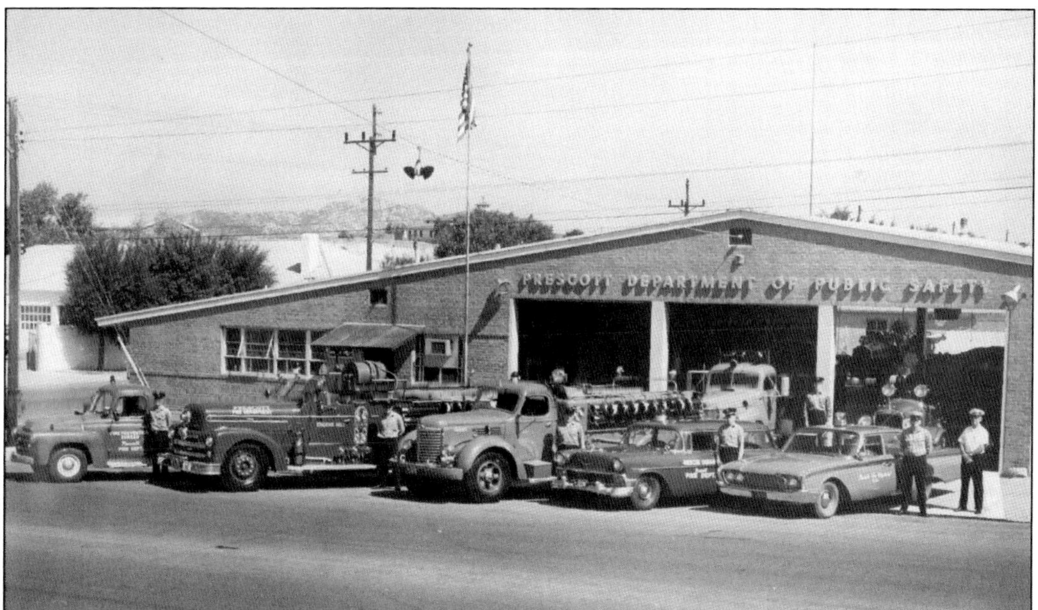

This 1960 photograph shows the better part of the fleet at the time. From left to right are a 1956 International used by the inspector, the 1957 Seagrave, the 1948 KB8 International with the 1945 KB5 International parked immediately behind it, a 1956 Chevrolet station wagon used as a rescue vehicle, and a 1960 Ford used by Chief Jack Custer. The 1931 Seagrave ladder truck is behind Custer's car, and the 1926 American LaFrance can be seen through the bays.

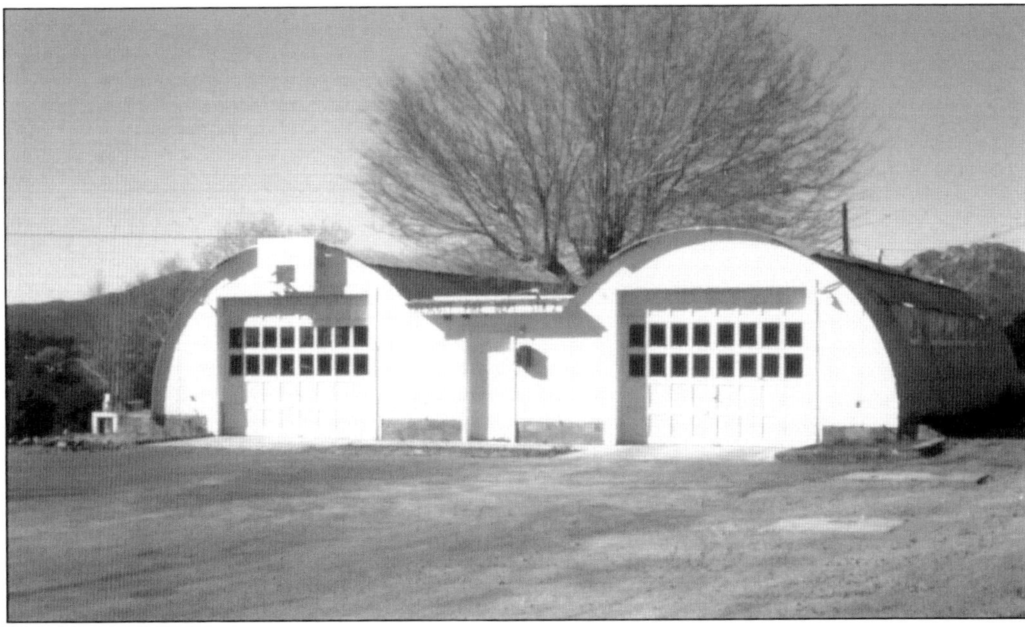

The Willow Creek Station in Prescott Heights, made from Quonset huts, was built in 1962 and opened in 1963. It served as an active station until 1979, when the opening of Station 72 on Iron Springs Road reduced its call volume to such an extent that it was frequently closed; its personnel worked out of the other stations in order to meet manning shortages there.

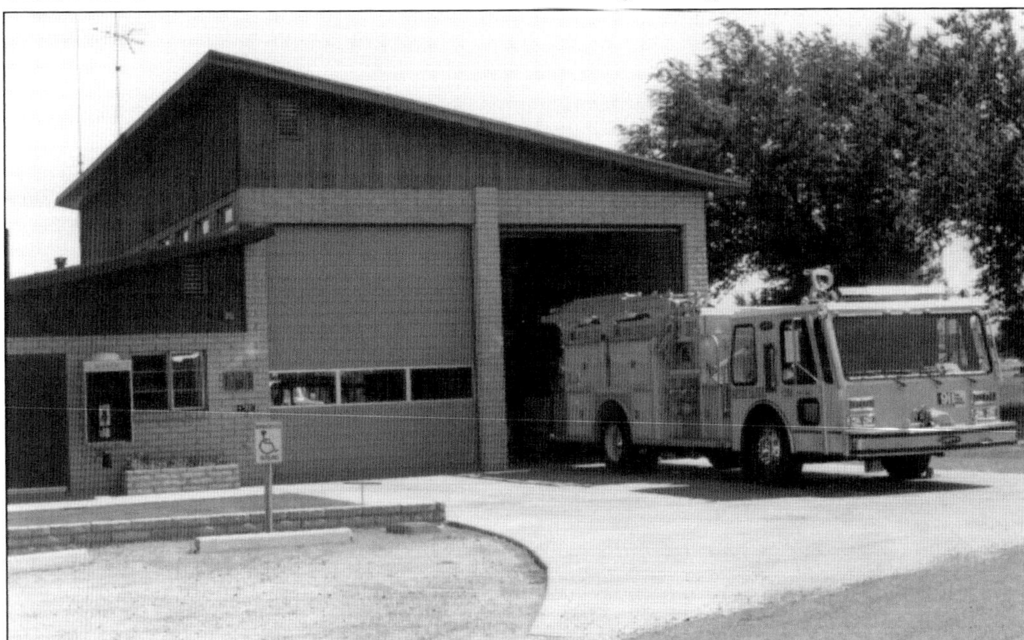

Pictured is Station 73 at 1980 Clubhouse Drive. This station was built in 1974, when Prescott's population was only 16,000. The station sits on the perimeter of the airport, with an engine and a patrol responding out of the front of the station and a crash vehicle responding onto the airport from the rear bay. The 1984 E-One engine, which has since been retired, is seen on the front apron. The E-One cost $137,000.

Built as Station 72 in 1978 at a cost of $543,842, this station initially housed dispatch and administration in addition to the engine company. Eventually, dispatch would move downtown to the regional dispatch center. In 2009, Central Yavapai Fire District took over the crew quarters and changed the name of the station to Station 51 while the Prescott Fire Department administrative and prevention staff continued to operate out of the building.

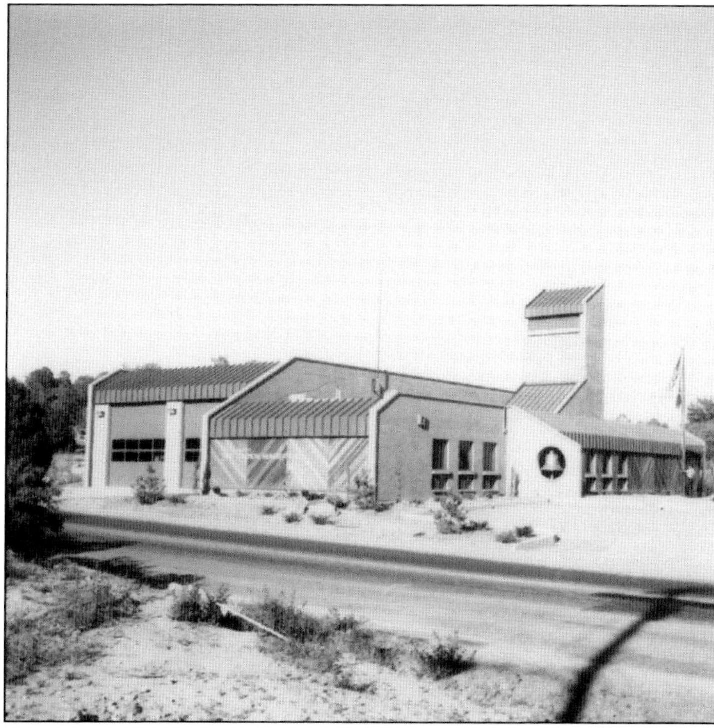

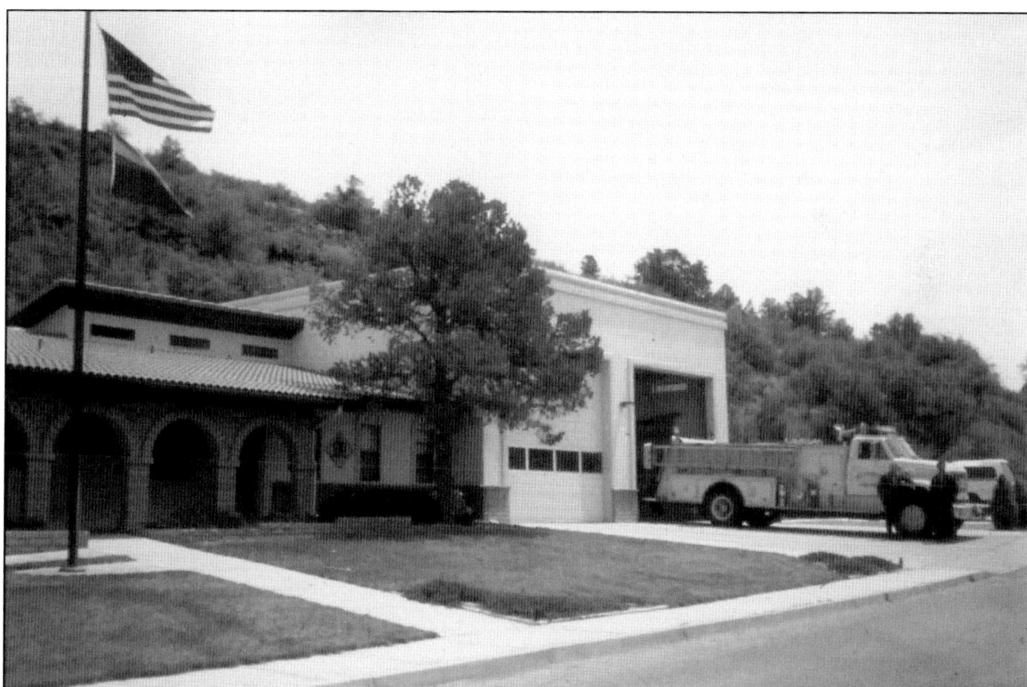

Station 74 began as a budget request for $539,000 in 1979, when the Quonset huts on Willow Creek were closed and a station farther north was needed. Requests for funding for Station 74 would continue until it was approved and built in 1987. The population in Prescott had grown to 25,000 by that point.

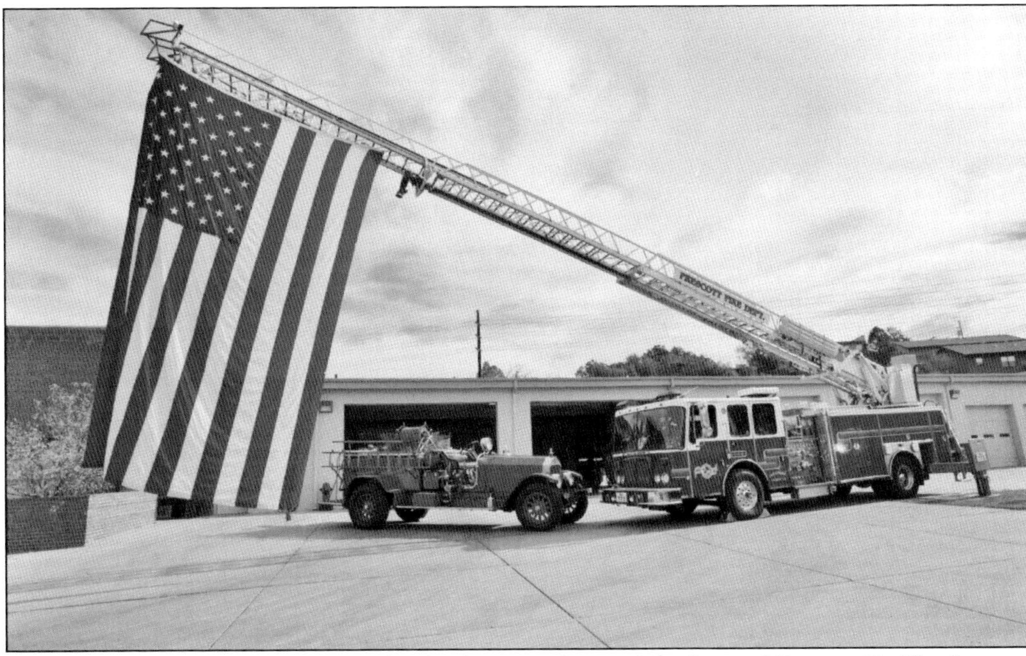

The Frog and Truck 71, nicknamed "Floyd," sit in front of the bays of Station 71. Floyd got its name from the slightly pink shade of red that was used to paint it. Purchased in 1989, Station 71 was formerly a car dealership before the department took over the building in 1991. The building was remodeled using the $235,000 generated when the city sold the old Station 71 on Goodwin Street. (Courtesy of Wade Ward.)

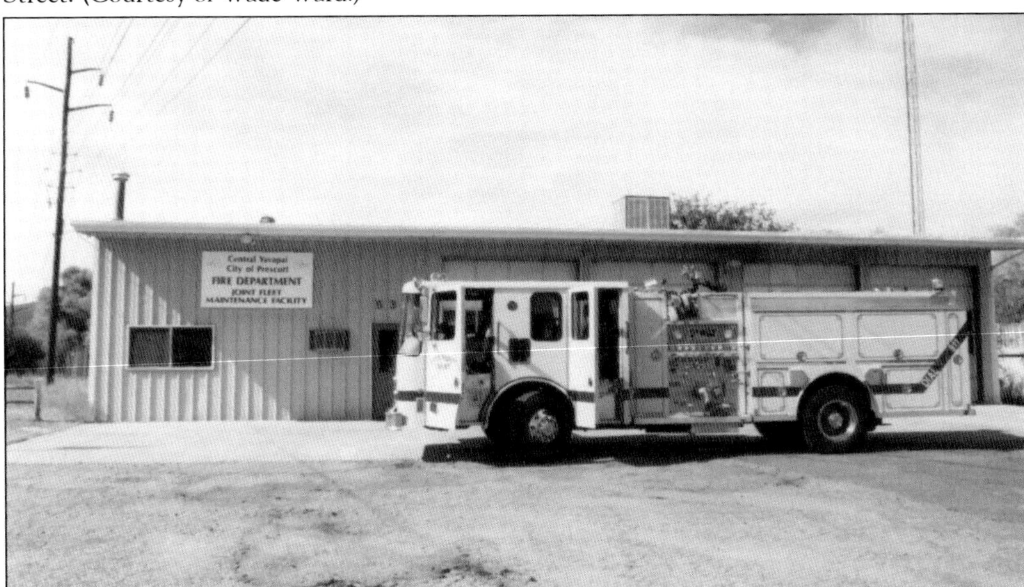

A building with a mixed identity, this Sixth Street structure was first built in the late 1960s and used as a mechanic's barn and station by Central Yavapai Fire District. In 1998, Prescott Fire Department would share the building with the mechanics, operating out of the building first as an adaptive response unit, and then as an engine in 2001, calling it Station 76. It was renamed Station 72 when Central Yavapai moved into the Iron Springs station in 2009.

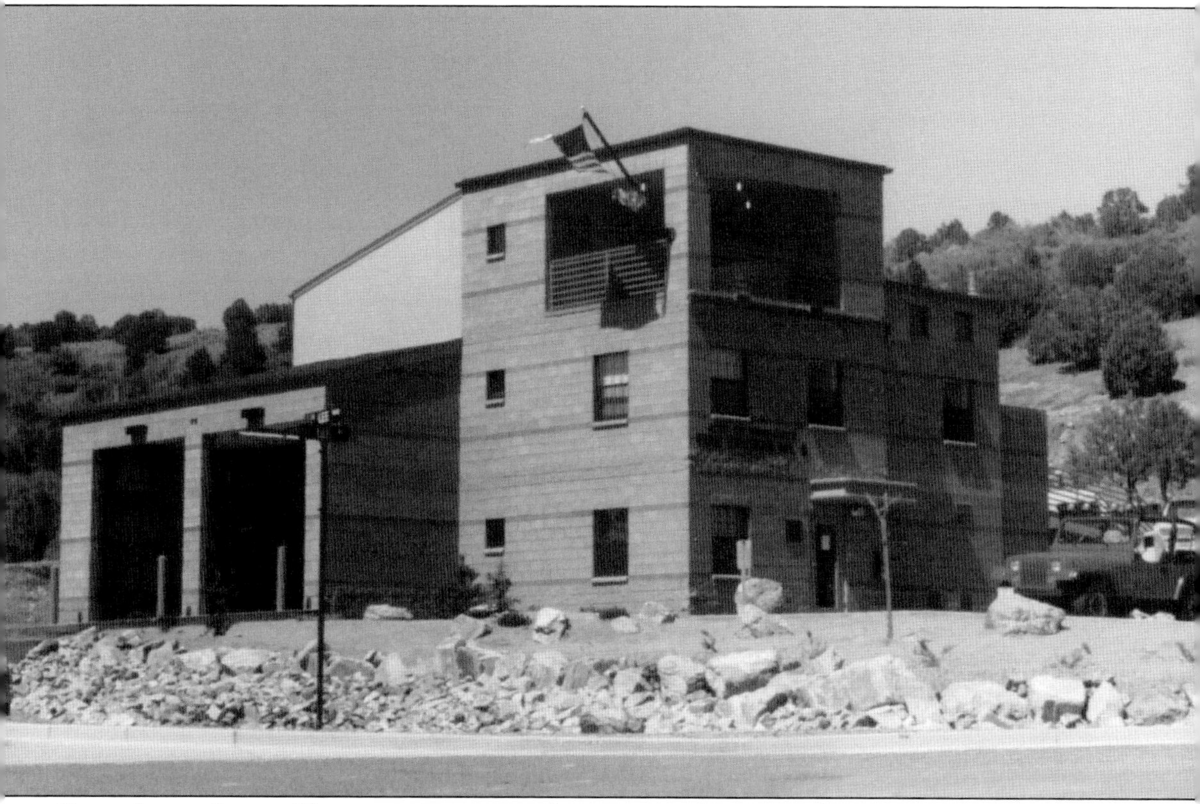

Shown here is Station 75 as seen when originally constructed in 1993, when Prescott's population was 31,000. Originally, it had only two bays and an open patio on the top floor. Central Yavapai Fire District mutually occupied the building with Prescott Fire Department on differing shifts as part of an agreement that kept hazardous material technicians from both departments at the station. Prescott Fire Department later moved into the building exclusively when Central Yavapai took over the Sixth Street station. The patio would eventually be walled in to make more office space. Two more bays and an additional crew quarters were added in 2001. The new crew quarters and one bay were subsequently leased out to Lifeline Ambulance. The station was originally designed to make use of a firepole, but the high cost of the pole resulted in its removal from the final product. Firefighters have to descend three flights of stairs to get from the sleeping quarters to the engine bay.

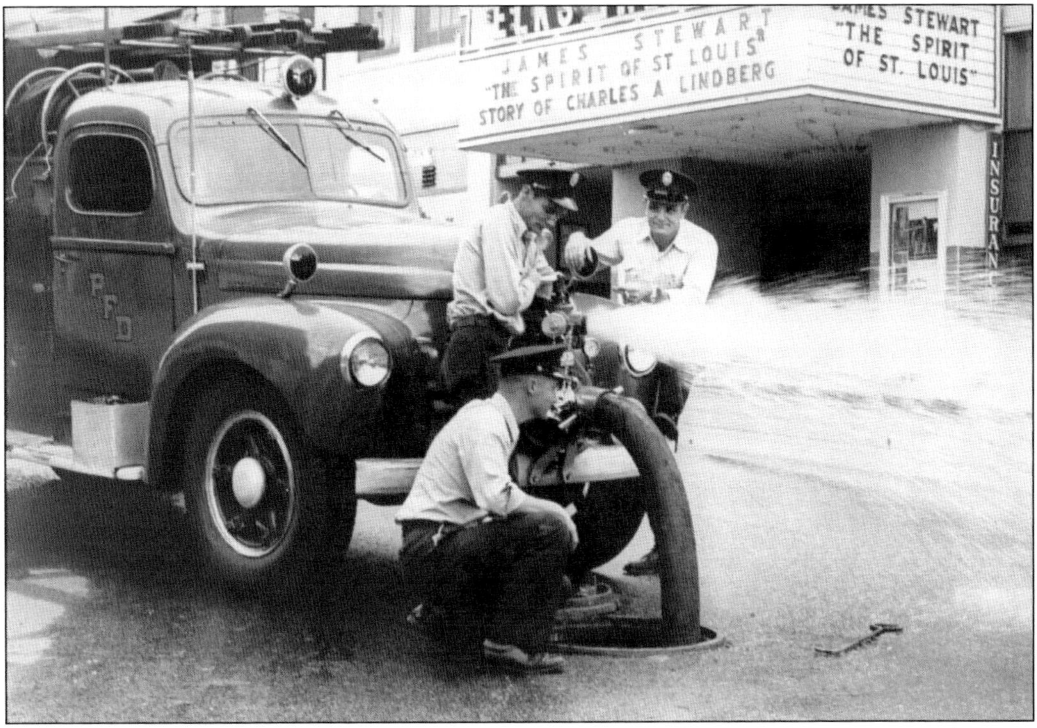

Ted Allen (left) and Vic Lira, with George Ravetto kneeling, use the 1945 KB5 International engine to clear water and debris from the storm drain in front of Elks Theatre. The KB5 was notorious for poor cornering, as its water tank lacked baffles to reduce the energy of the shifting water.

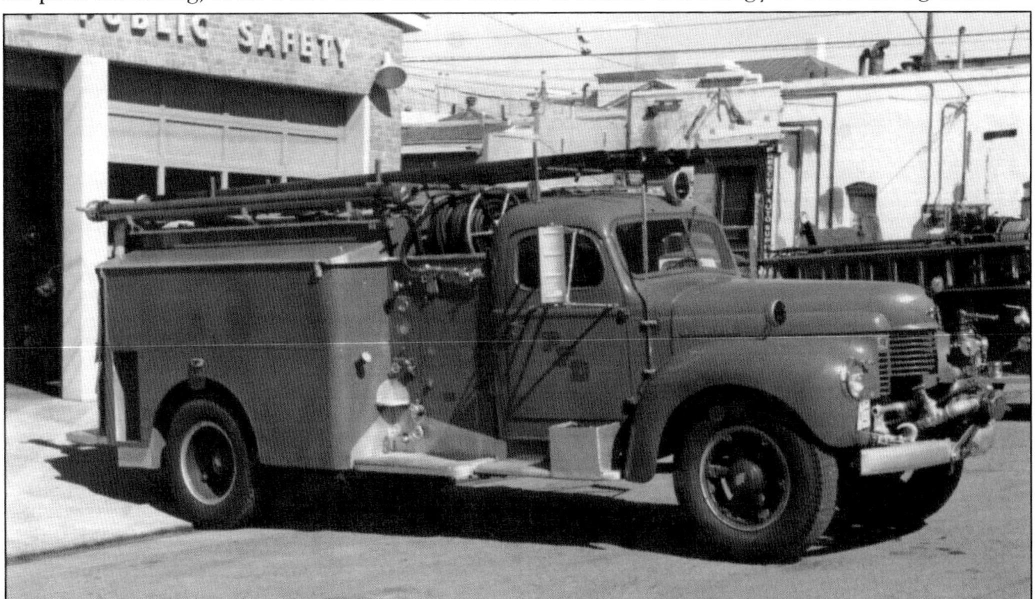

The 1945 KB5 was a military surplus vehicle originally purchased by the city's streets department and found to be ineffective. The vehicle was transferred to the fire department and a Hale pump was mounted to the front. The weight of the water tank and equipment in the rear of the vehicle reportedly caused the front of the truck to lift off the ground.

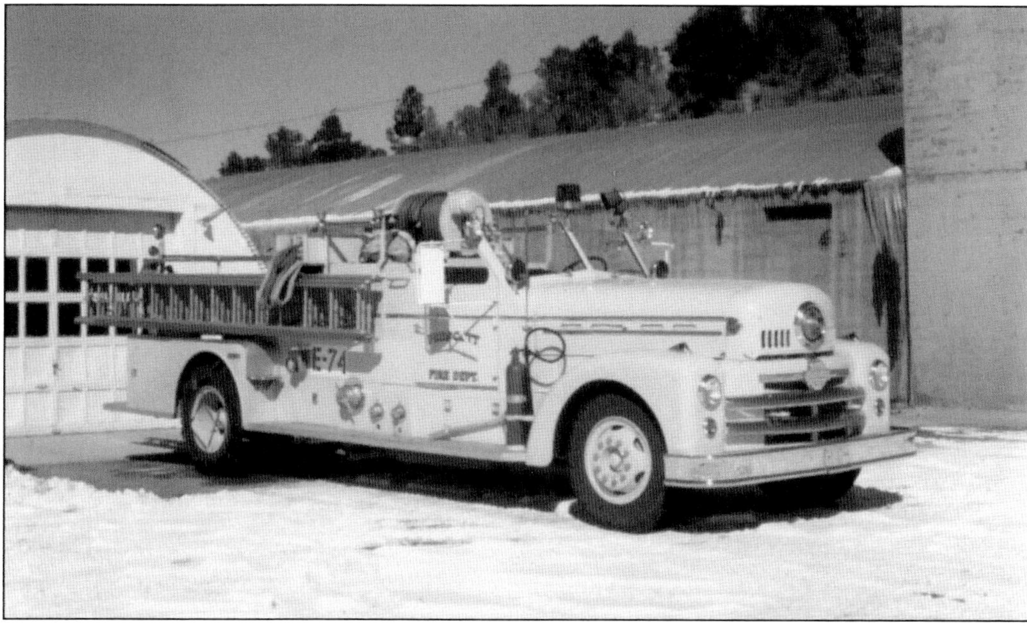

Seen here after its new yellow paint job is the 1957 Seagrave triple combination pumper. The vehicle was purchased for $28,711. It had a 1,250-gallon-per-minute pump and a 250-foot hose reel. The council was debating three different engines when a self-described "peddler" for Seagrave made arguments in the Seagrave's favor, resulting in a winning sale. The department still owns and stores this engine for future use as a parade vehicle.

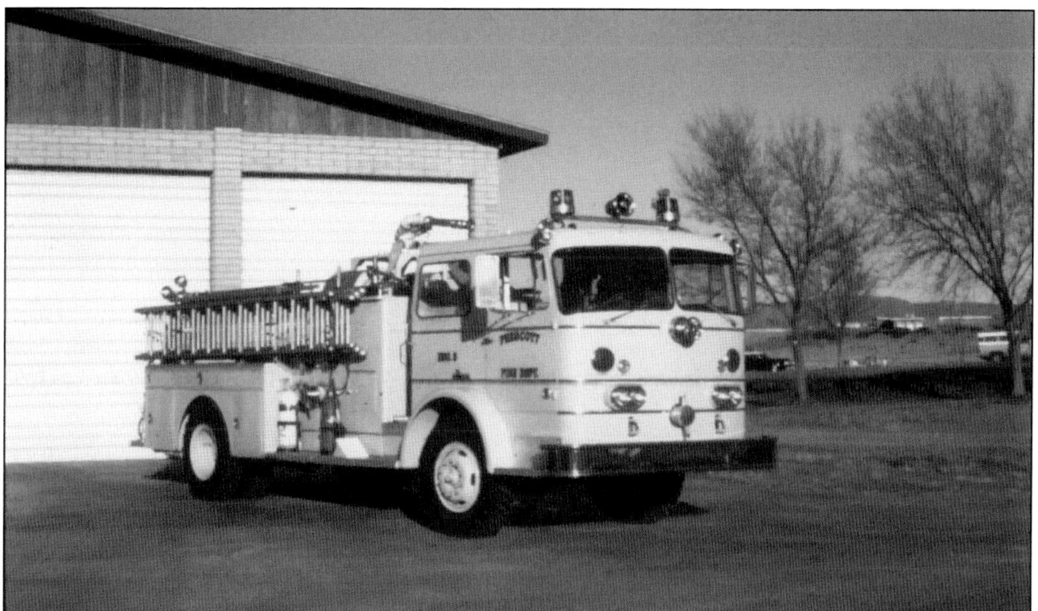

In 1963, two 1,250-gallon-per-minute FWD pumpers were purchased for $28,770 each. The purchase was in response to threats of increasing insurance premiums as the department's fleet was aging. A novel feature of the pumpers was the closed cab, the first owned by the department. Seen here in 1977, this FWD has received a fresh coat of paint, going from the original red to the yellow of the 1970s.

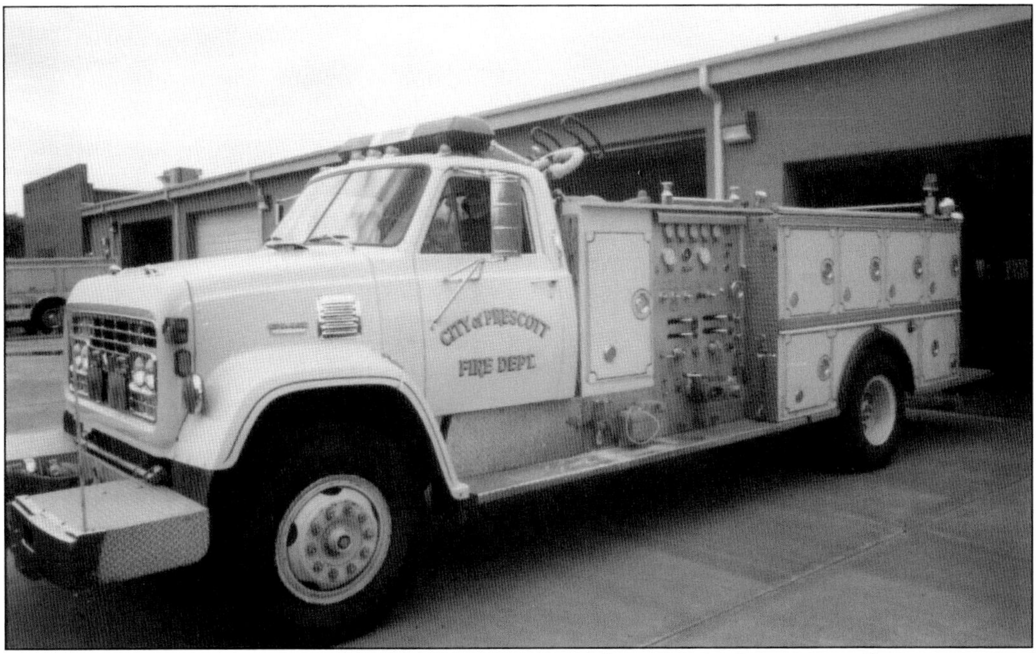

Arriving in March 1975, this GMC pumper cost only $48,000, had a 1,000-gallon-per-minute pump, and was one of two GMC pumpers that were purchased. The arrival of the new equipment allowed the department to sell its 1948 KB8 to Prescott's sister city of Caborca, Mexico, for $1. The two GMCs were eventually sold to Grand Canyon and Tusayan Fire Departments.

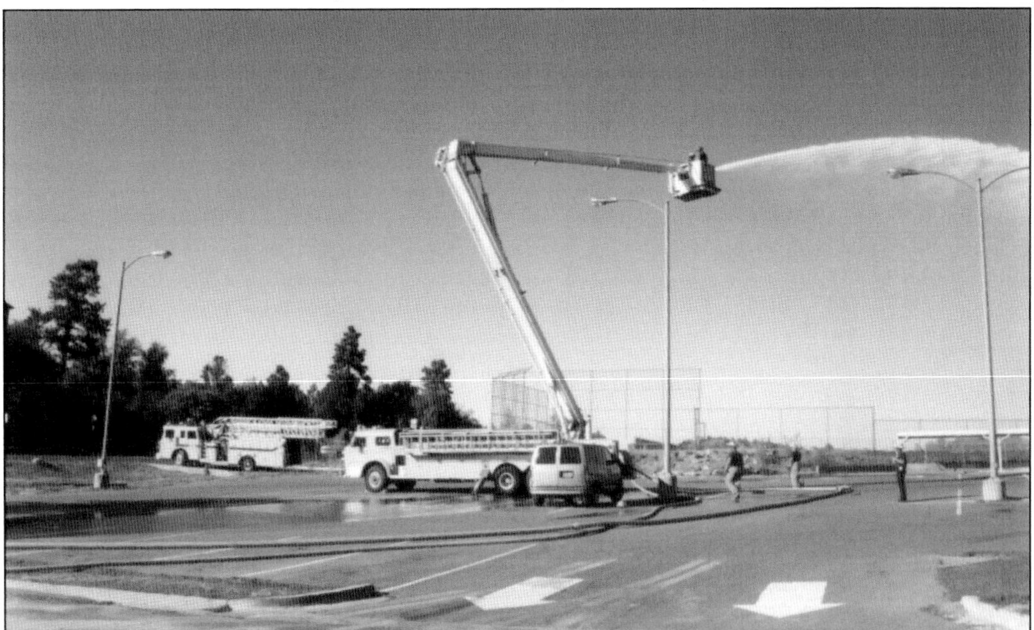

The 85-foot American LaFrance Snorkel was delivered in March 1975, six months overdue, at a cost of $94,500. The Snorkel was too tall to fit into the station on Goodwin Street, so the farthest bay door had to be enlarged. The Snorkel lacked a pump, and its aerial nozzle had to be pumped by another engine.

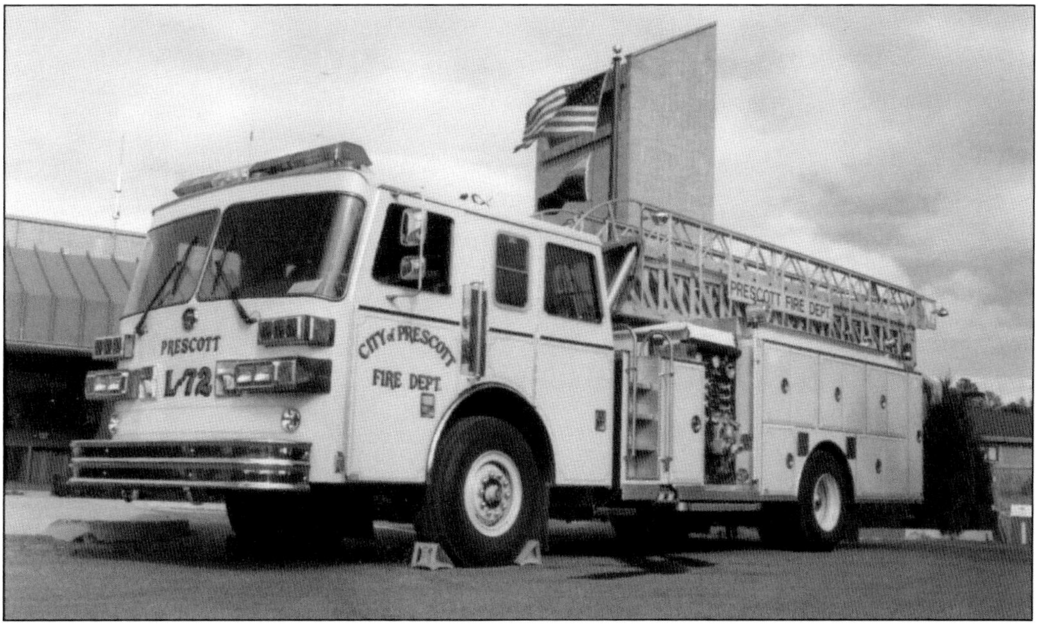

The Sutphen Ladder was a 1,500-gallon-per-minute quint, or pumping ladder truck, purchased in 1989 at a cost of $301,000. It was a mid-ship platform with a 75-foot ladder. Its delivery was much delayed, earning it the nickname of the "Myth." Its name was scribed minutely in the blue trim paint around the window of the engineer's door. Many probationary firefighters were tasked with finding the diminutive script.

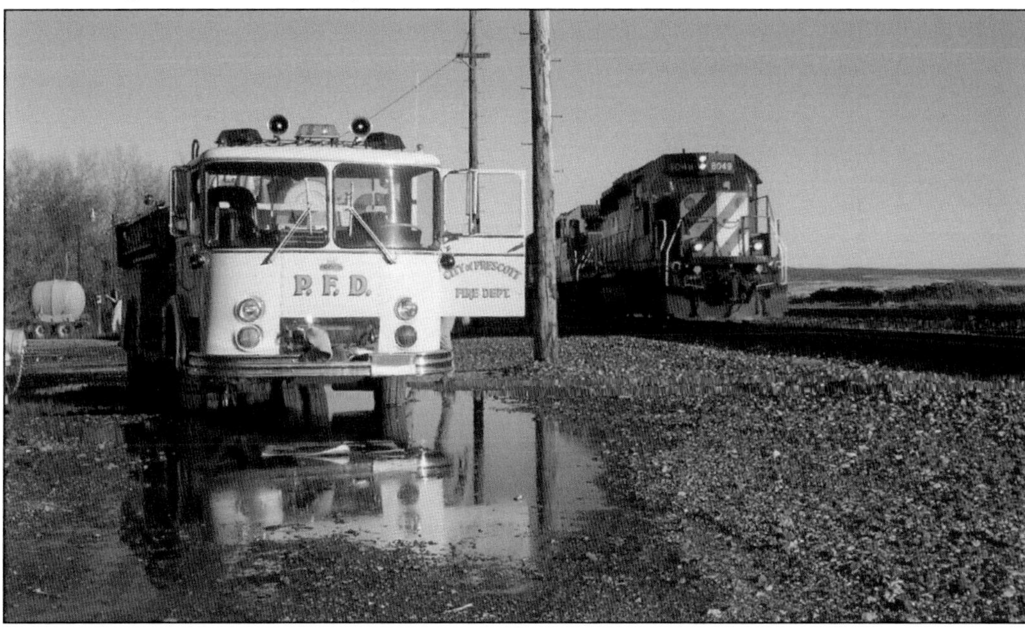

The 1979 Crown fire engine was purchased for $84,000. It had a 1,500-gallon-per-minute pump. Originally priced at $103,000, the Crown company dropped the price in order to be able to deliver a third Crown pumper to Arizona, with the other two going to Tucson Fire Department. The Crown would eventually be retired and loaned to Prescott High School in 2000 for its fire science program and then sold to Seligman Fire Department in 2005 for $1.

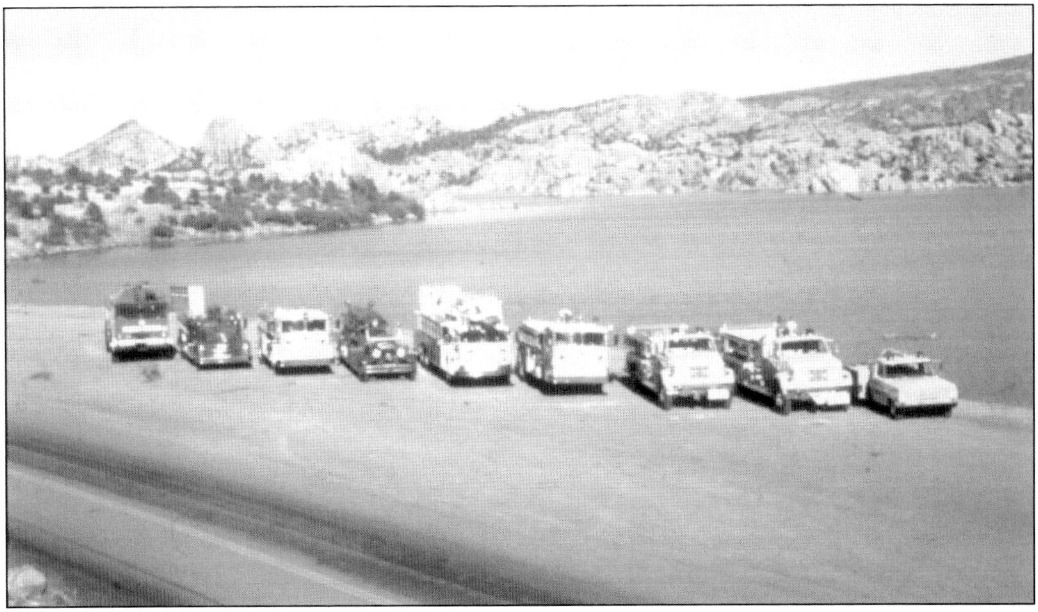

The size and color of the fleet changed over the years, starting as red, then turning yellow in the 1970s, and returning to the traditional red in the 1990s. Seen here is the fleet of engines in 1975 and again in 2007. The 1975 photograph has two GMCs, two FWDs, two Seagraves, the Snorkel, an aircraft crash engine, and a custom-made patrol unit. By 2007, the fleet had nearly doubled in size and included two aerials, four Type-One engines, two Type-Three engines, two Type-Six wildland patrols, a technical rescue vehicle with a trench rescue trailer and a boat, a hazardous materials vehicle, an E-One aircraft engine, and two crew buggies for the Granite Mountain Hotshots.

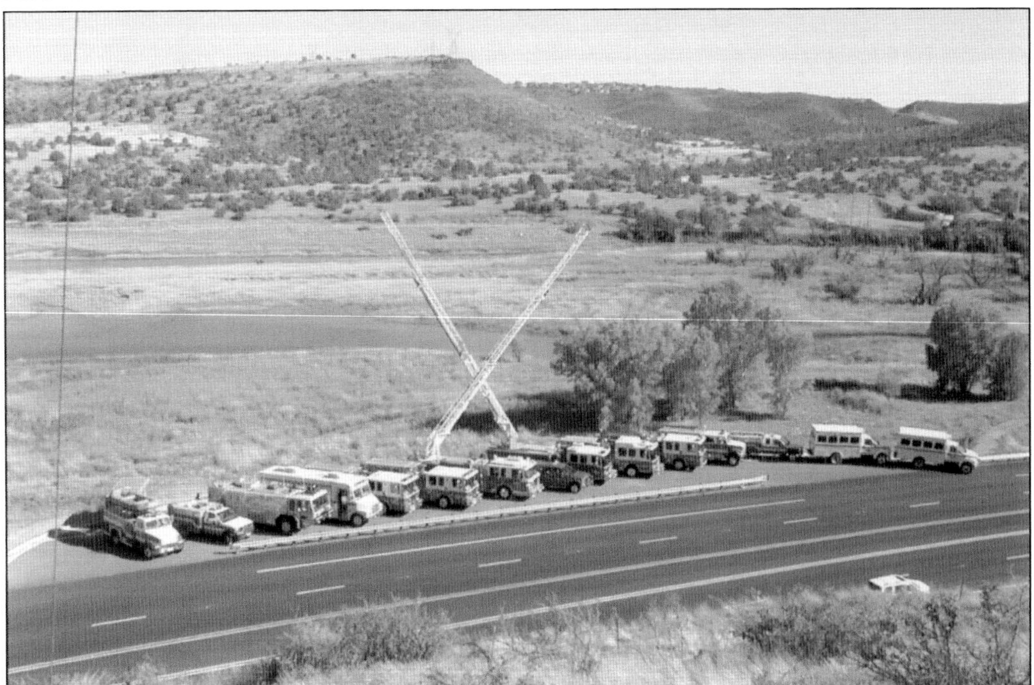

Four

Firefighters and Chiefs

In Prescott Fire Department's 128-year history, it has seen the leadership of 30 chiefs. While each of the four original companies met monthly and selected their own foremen and assistant foremen, the whole of the department would meet four times annually; at the December meeting, they would select a chief and first and second chief, as well as a secretary and a treasurer from among the ranks. Their length of tenure has varied. Early on, chiefs served for two years, with the exception of the first chief, Al Whitney, who failed to complete a full year of service before being replaced by Charles Spencer.

After the fire of 1900, their lengths of service expanded, with Anthony "Tony" Johns serving for 14 years after the Great Fire. The practice of the firefighters selecting their own chief would continue until 1956, when the city determined that it needed to select the chief and make it a paid position. The cause for the change met the demands of the National Board of Fire Underwriters and needed to occur to save the citizens money on their insurance billing. That year, the city selected Jack Custer from a pool of five applicants. Custer served 18 years with the department as chief, the longest of any of the chiefs.

Staffing would see considerable change over the century. The 100 volunteers were amended with two paid positions in 1906, which would remain the standard until the 1950s, when insurance rates would begin to affect the citizens. The transition was rapid. Within 20 years the department grew to 30 paid positions, and in 1975, the volunteer contingent finally disappeared. Paid firefighters numbered around 40 through the 1980s and eventually climbed to nearly 60 at the turn of the 21st century. Today, including the wildland division, the Prescott Fire Department is roughly 80 strong.

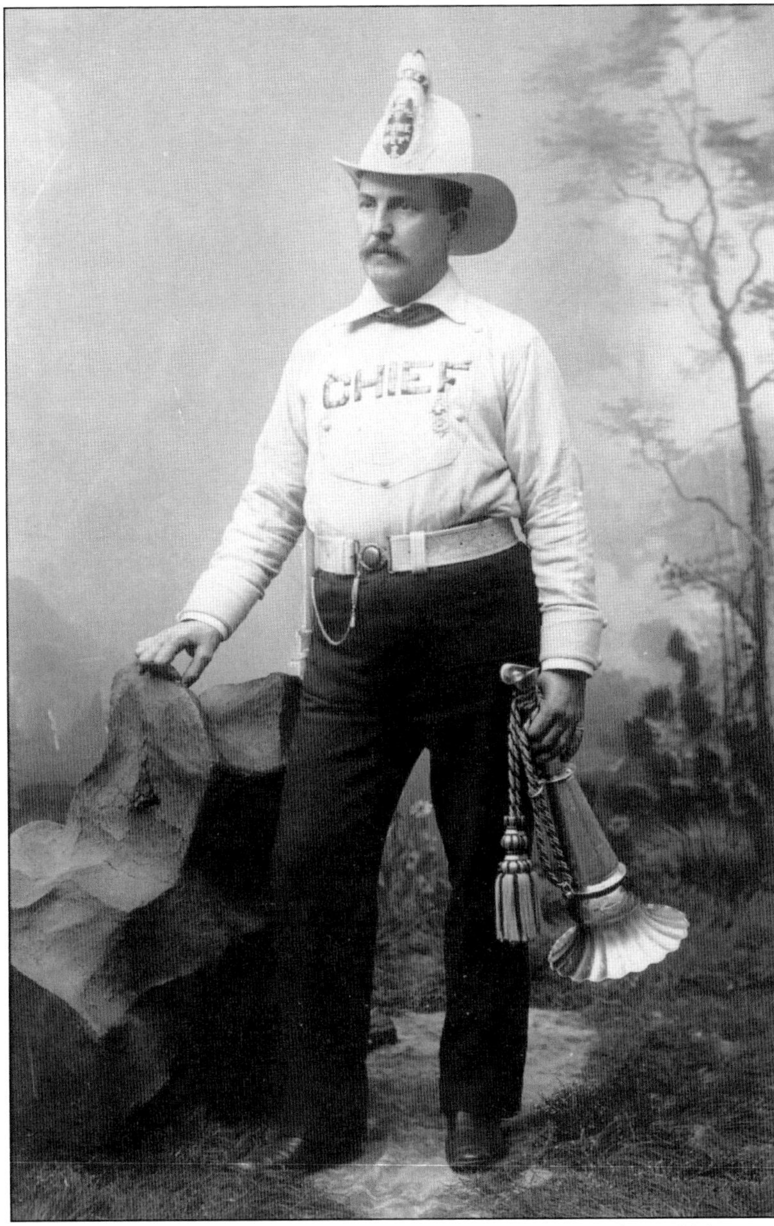

Will Coles Bashford was the Toughs foreman at their inception and would later take the reins as chief in 1889. In 1891, Mayor John Howard and Chief Bashford had a very public argument in the *Journal-Miner* newspaper. Bashford had attempted to conduct a readiness drill by ringing the fire bell in the courthouse. Howard ordered the ringing stopped and thus started a debate of who was ultimately in charge of the bell. Howard argued that the citizens could not "afford to be inconvenienced by having the fire bell rung" without an actual fire. Bashford's retort stated that the bell was in fact an appliance of the department, of which he had authority and, "Whenever I feel that it is my duty . . . I shall order the bell rung!" Bashford ultimately won the argument as the county's board of supervisors resolved the following week that while the bell was under the charge of the sheriff, the chief could use it as he saw fit in the best interests of the department. (Courtesy of Sharlot Hall Museum, #po911p.)

Anthony "Tony" Johns served as chief from 1901 to 1915. Born in England in 1864, he moved to Prescott in 1882 and lived at 105 Grove Avenue. He served as undersheriff from 1898 to 1902. After his time as chief, Johns served in the state legislature as speaker of the house from 1917 to 1918 and president of the senate from 1918 to 1919. (Courtesy of Sharlot Hall Museum, #po0868.1pb.)

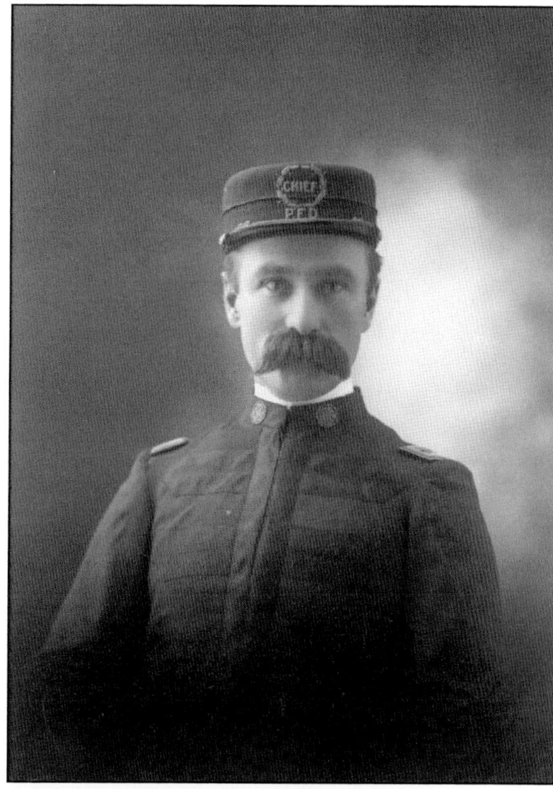

At the bottom left of this c. 1915 photograph is Bert Tilton, who served as chief from 1915 to 1920. Tilton came to Prescott in 1893 at age 21 with only $27.30 in his pocket. His estate was valued at a quarter of a million dollars at the time of his death in 1940. Tilton died of a heart attack brought on after his music store was consumed in a fire. (Courtesy of Sharlot Hall Museum, #fi103p.)

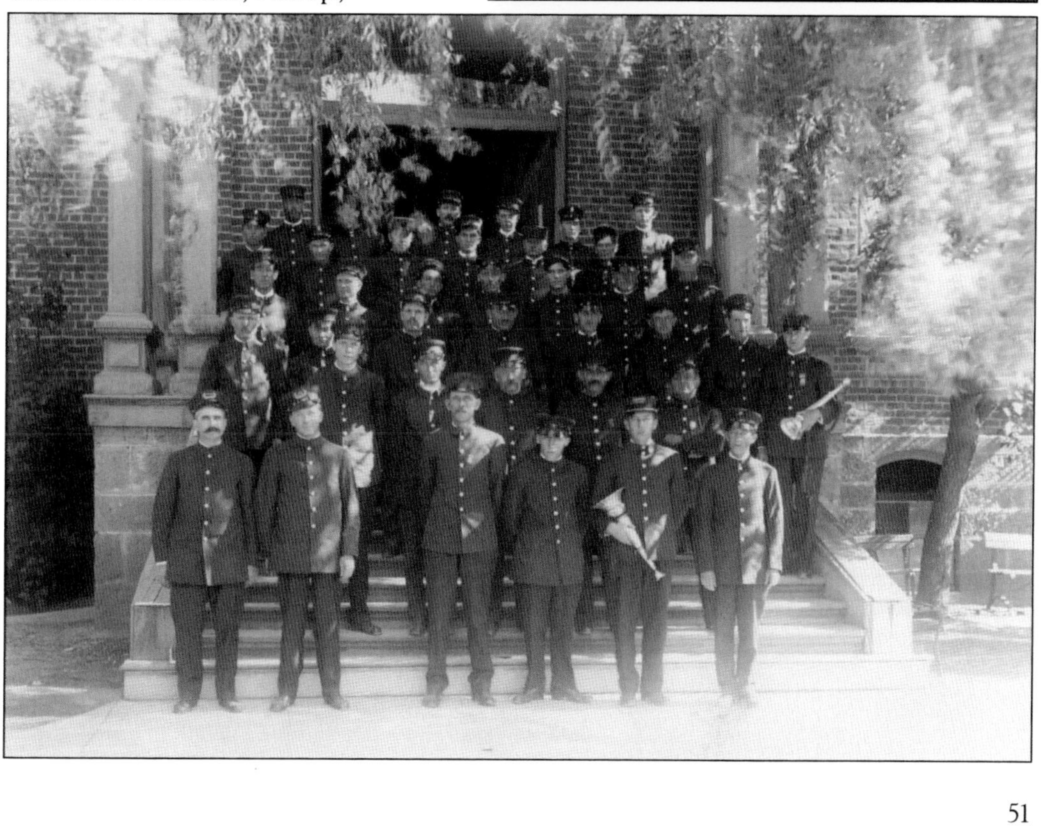

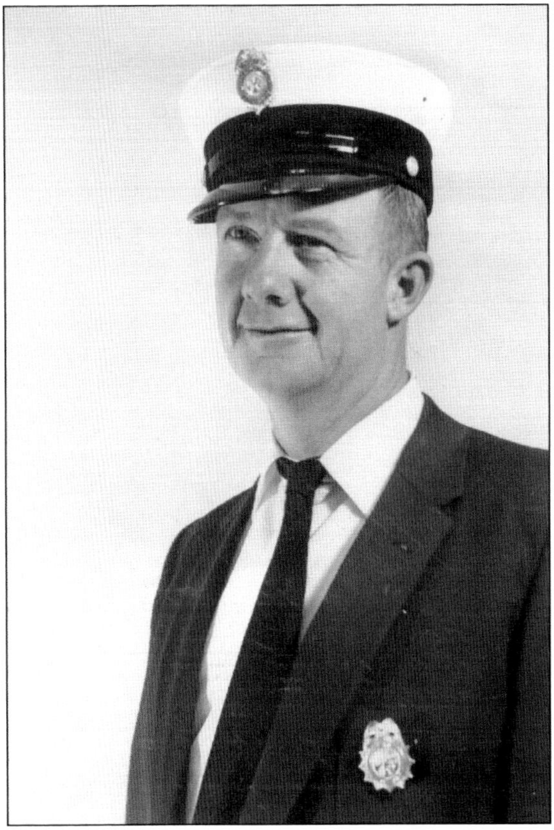

Enoch Seale joined the department when the Velie and Model Ts were still in use and saw the purchase of the Frog and the Seagrave truck. He served as chief from 1952 to 1954. He would later serve as assistant chief under Custer after Hawke retired. Seale served 23 years on the department.

Seen here from left to right in 1956 are the original paid crew: Tony Oriet, Jerry Meiner, Gene Shimonowsky, Jack Custer, George Ravetto, Ted Allen, and Joe Mackin. Each was given a specific function: Oriet was in charge of hydrants; Meiner, prevention; Shimonowsky, training; Ravetto as a "filler" where needed; Allen, fire alarms; and Mackin, extinguishers and record keeping. (Photograph by Wade Corley.)

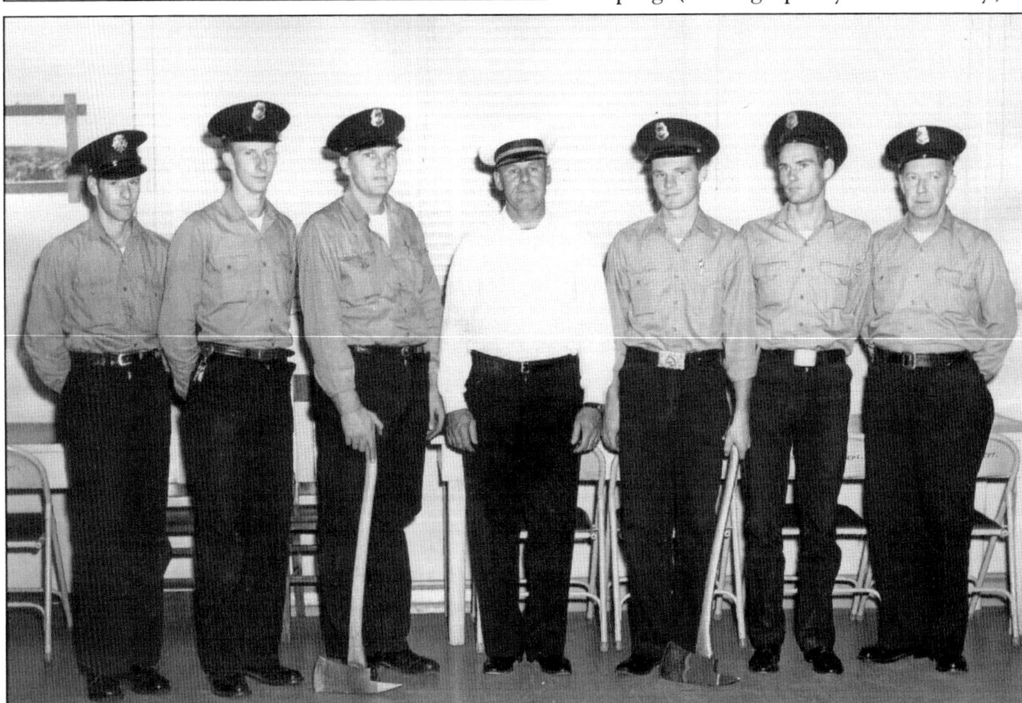

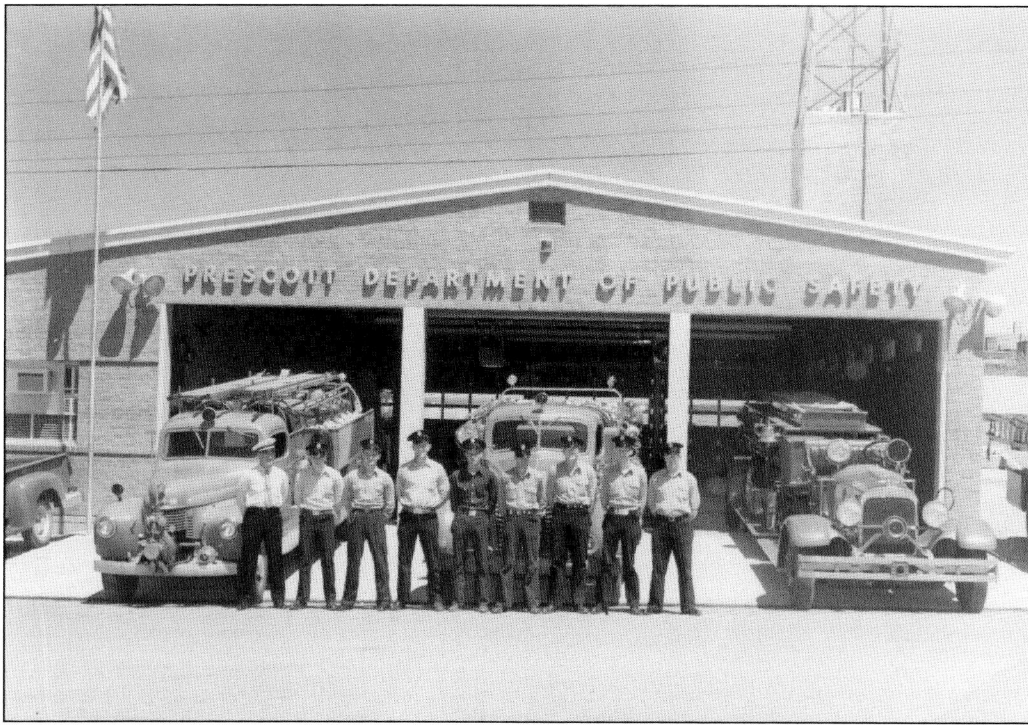

By 1957, the department had grown to nine paid firefighters. The move to a paid department was heavily influenced by the city's insurance evaluations. Shown here from left to right are Jack Custer, Vic Lira, George Ravetto, Josh Spivey, George Allen, Tony Oriet, Gene Shimonowsky, Jerry Meiner, and Joe Mackin. New firefighters were paid $288 per month, with engineers receiving $390.

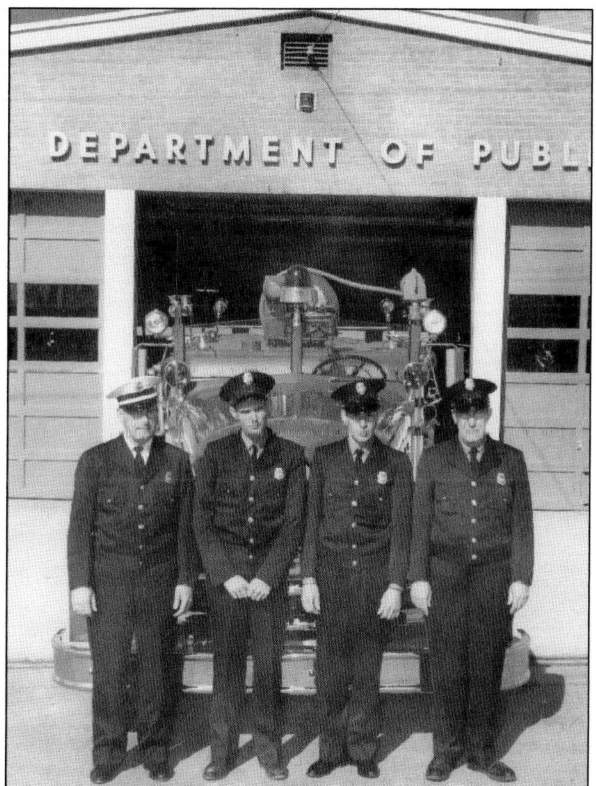

From left to right in this 1958 photograph, Jack Custer, Ted Allen, Tony Oriet, and Joe Mackin were the newest hires to the department, wearing the new department uniforms. The new uniforms were purchased because the firefighters were conducting home inspections and there was concern that they should be readily identifiable to the public as firefighters.

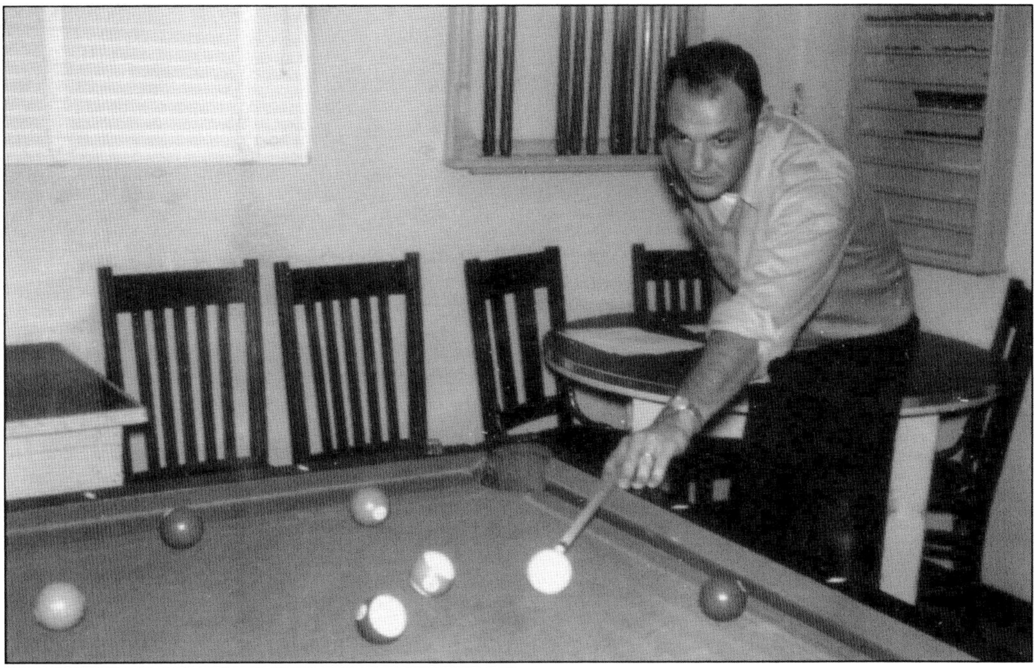

Vic Lira is playing on one of the department's two Brunswick billiard tables. The tables were purchased in 1915 with the justification that their presence would give the firefighters something to do and "keep the young fellows from getting into trouble." One of the tables was later donated to the Arizona Pioneers Home in 1954 prior to moving into the Goodwin Street station. The remaining table currently resides at Station 71. Department legend has it that Brunswick offered to furnish a top end table to every station in exchange for this table, one of their first ball-return tables. Below, Lira shows students around the station and engine. Vic was hired in the spring of 1957 at a salary of $288 per month. After only two months with the department, he won the hose-rolling contest at the state convention in Prescott that summer.

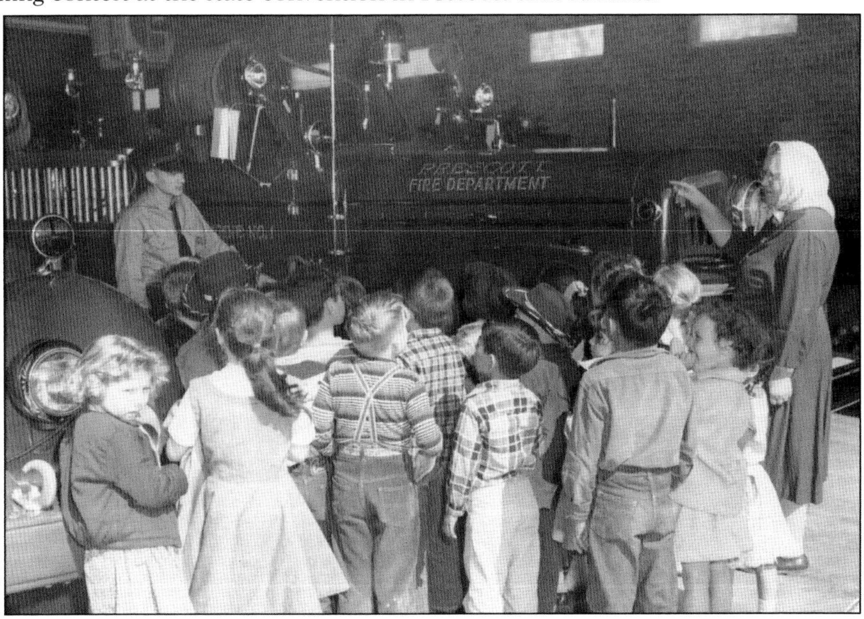

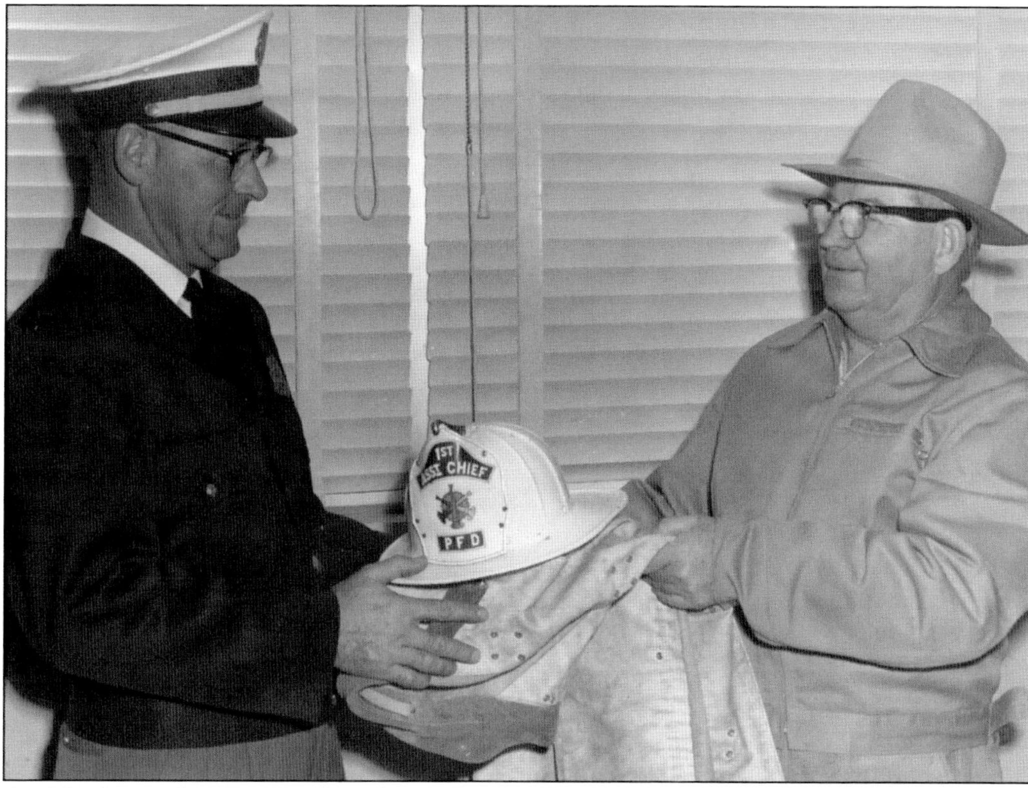

Art Hawke is submitting his formal resignation to Chief Custer. Hawke served as chief from 1954 to 1956 and was the last of the chiefs chosen by the membership. After his term as chief, Hawke served as assistant chief under Custer until 1959. Hawke oversaw the department's transition in 1954 from the Prescott Volunteer Fire Department and four companies to one Prescott Fire Department. Thirty-five of the younger men were selected as regular volunteers, and the remaining older men became the Prescott Firemen Association in support of the department. Three fulltime paid engineers were added at that time, bringing the number of paid firefighters to five.

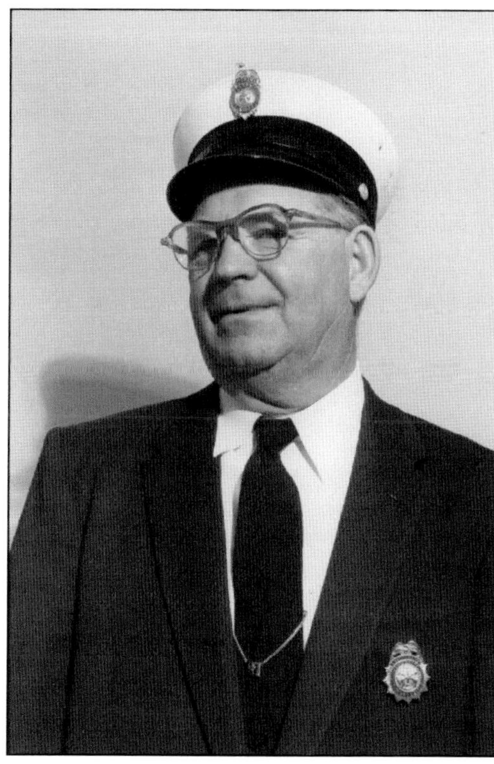

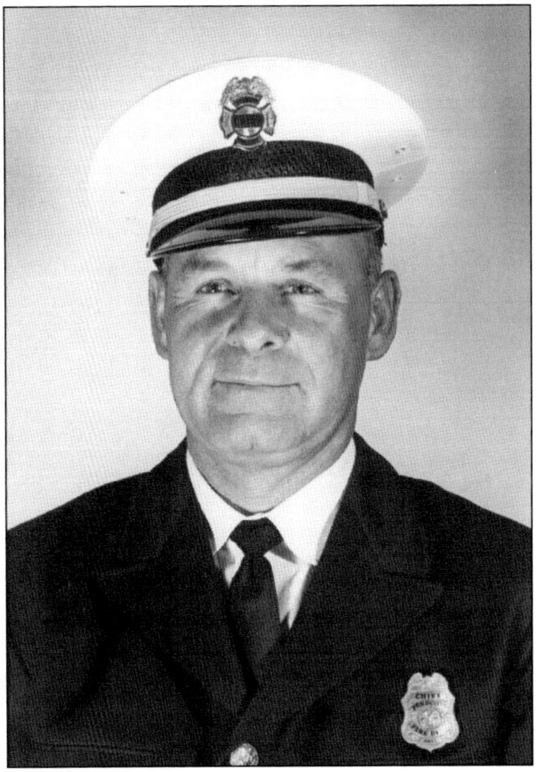

In 1956, Jack Custer had the distinction of being the department's first paid fulltime chief and the first chief selected by the city and not by the membership. Custer was originally a firefighter for Phoenix Fire Department in the 1930s and 1940s before becoming the chief at the Whipple Veteran's Administration in Prescott and then chief for Prescott Fire Department. His pay was $420 per month. Custer was Prescott Fire Department's longest serving chief at 18 years.

The firefighters are gathered in front of the Frog in this 1957 photograph. From left to right are (first row) Second Assistant Chief Enoch Seale and Mac Reininger; (second row) Chief Jack Custer, fireman Vic Lira, engineer Tony Oriet, fireman Josh Spivey, Fire Inspector George Ravetto, fireman David Long, engineer Gene Shimonowsky, Capt. Doug Brown, Capt. Warren Smith, fireman Gordon Nelson, fireman John Bouton, and engineer Joe Mackin. Fireman Bob Eddings is in the driver's seat.

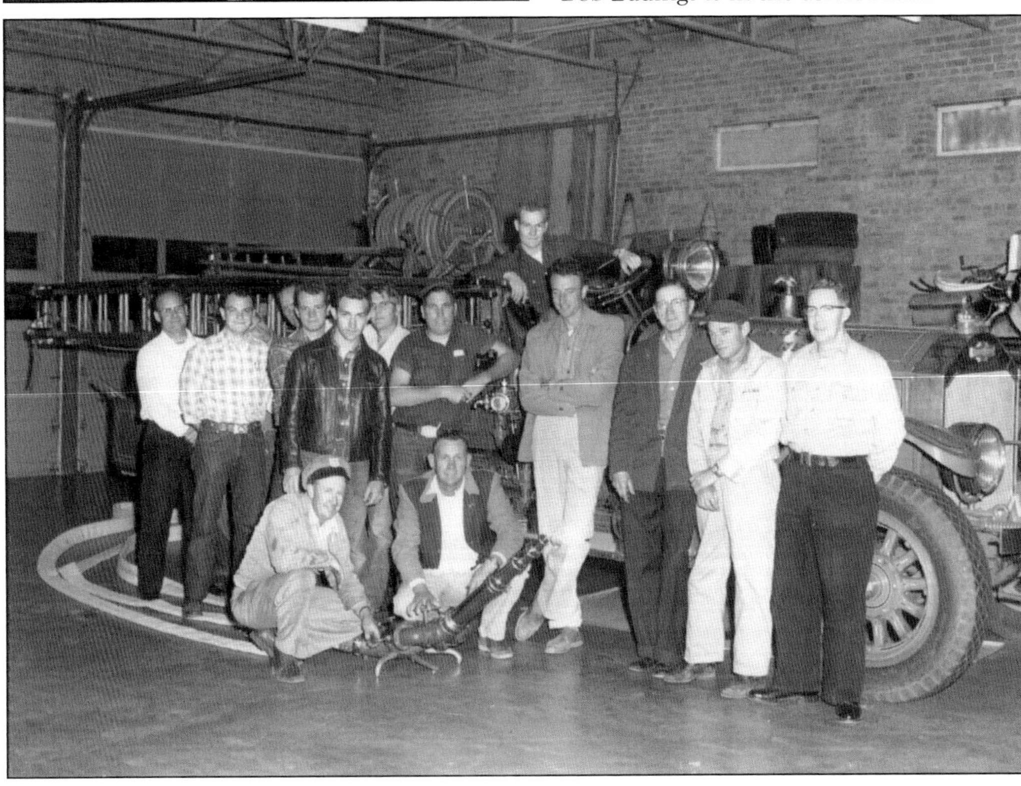

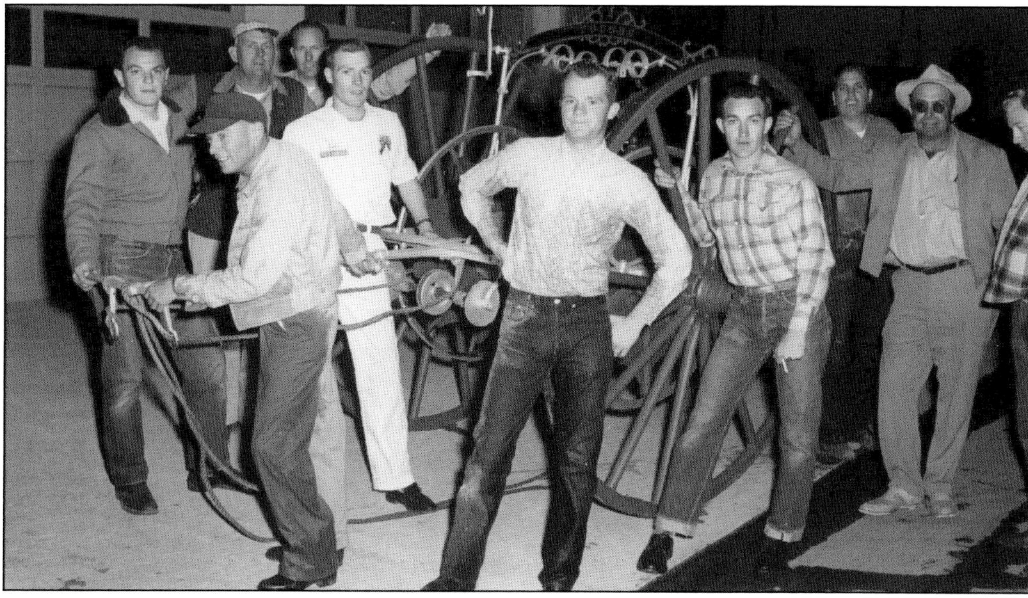

The hose-cart racing tradition had become lost over the years. The Prescott Fire Department hosted the 31st Annual Arizona State Firemen's Association convention in 1957, and all the departments attending were advised to dust off their hose carts and bring them to the convention for races. John Bouton pulls the cart and is followed by, from left to right, Bob Leckliter, George Ravetto, unidentified, Douglas Brown, Sid Webb, and George Hill. In the back left are Vic Lira, Mac Reininger, and Josh Spivey.

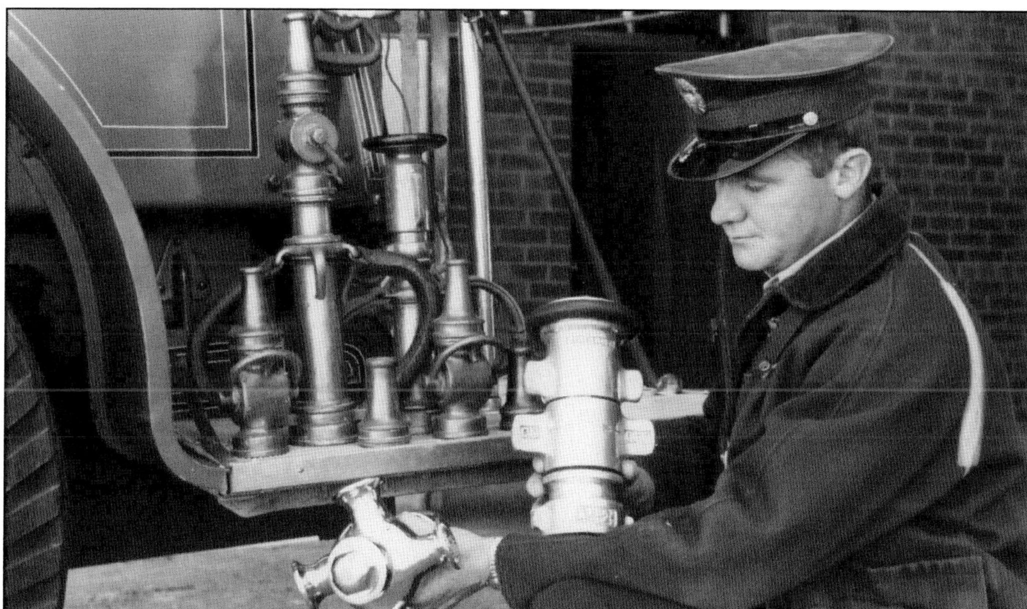

In this 1961 image, George Ravetto is seen inspecting the nozzles that were mounted on the tailboard on the 1931 Seagrave truck. Ravetto was originally hired to serve as an "as needed" firefighter, an "other duties as assigned" position that assisted the better defined positions such as training, mechanic, hydrants, records, extinguishers, and inspector. He would later take on the role of inspector.

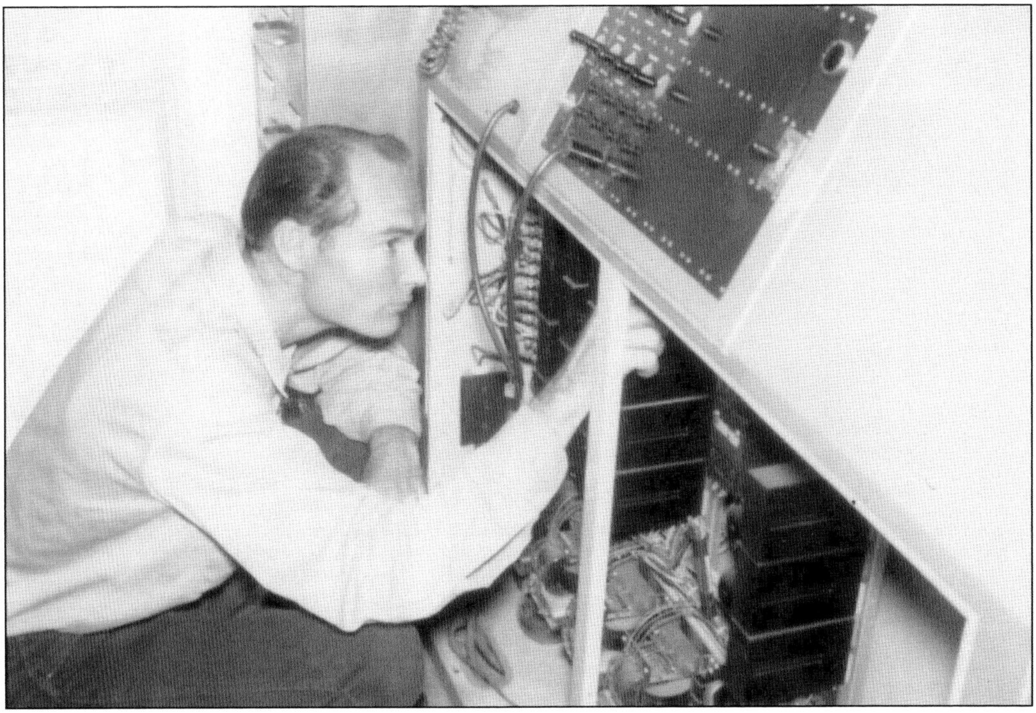

Ted Allen is conducting maintenance on the radio and alarm equipment. As is common practice even to this day, the firefighters each hold responsibility for managing different programs within the department. This may include addressing simple station needs like soap and mops, to highly specialized work like writing the detailed construction order for a half-million-dollar fire engine. Allen oversaw the update of the signal system from 80 to 215 Gamewell signal boxes across town. Below is the dispatch console at the Goodwin station. Emergency dispatches for both fire and police were handled from this location.

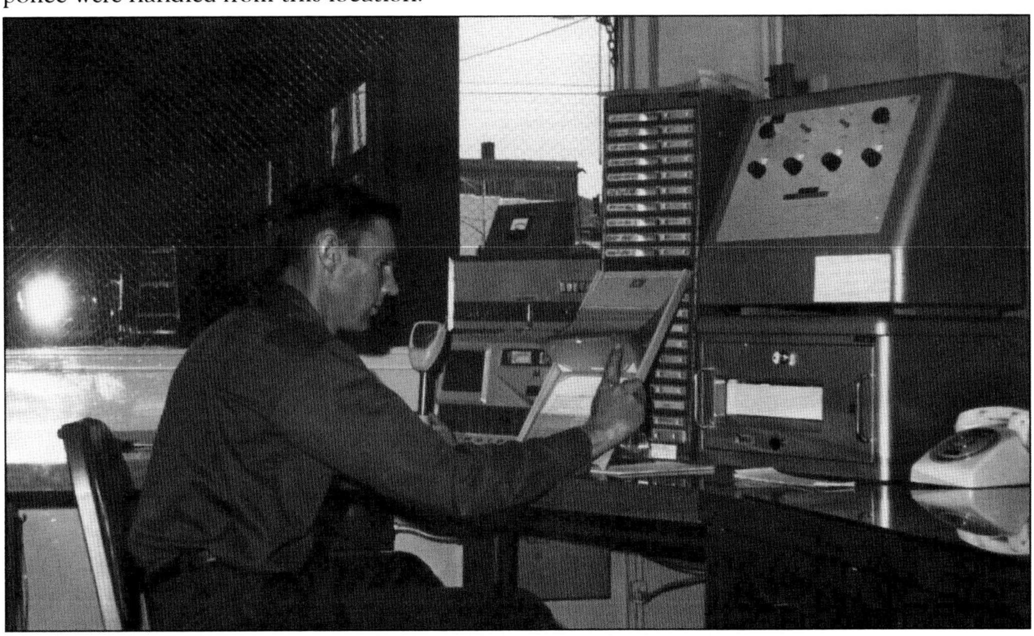

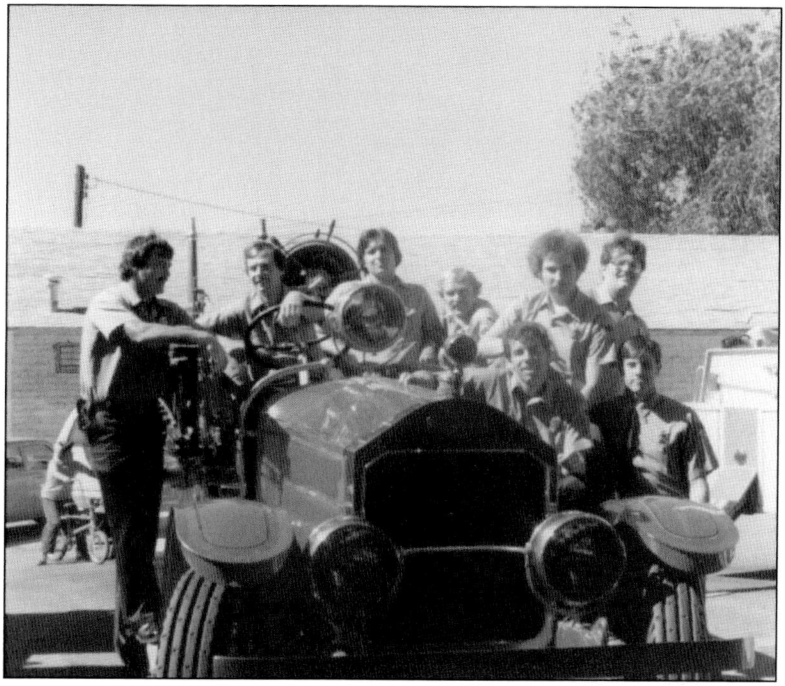

The rookie class of 1979 is seen here on the Frog. From left to right are Terry Sutton, Mike Parrish, Jim Mason, Tim McElwee, Brian Martinez, Virgil Esmont, John Bostwick, and Monty Farrar. The eight firefighters completed a four-week training academy prior to coming to the floor.

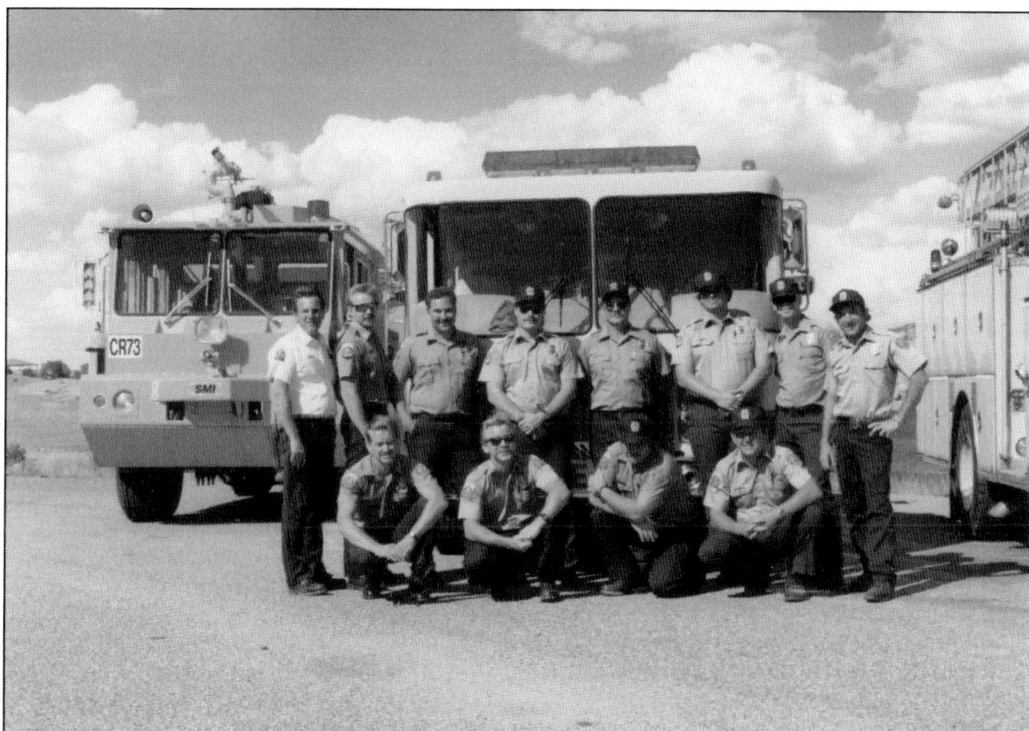

This photograph of C-shift was taken during a shot of the entire fleet at Watson Lake. From left to right are (first row) Dean Hinshaw, Scott "Kermit" MacDermid, Dan Sanders, and Jim Barber; (second row) Duane Steinbrink, Jeff Knotek, Virgil Esmont, Dennis Jeffrey, Perry Hickman, Jim Mason, Tim Sheehan, and Ed Hoffman.

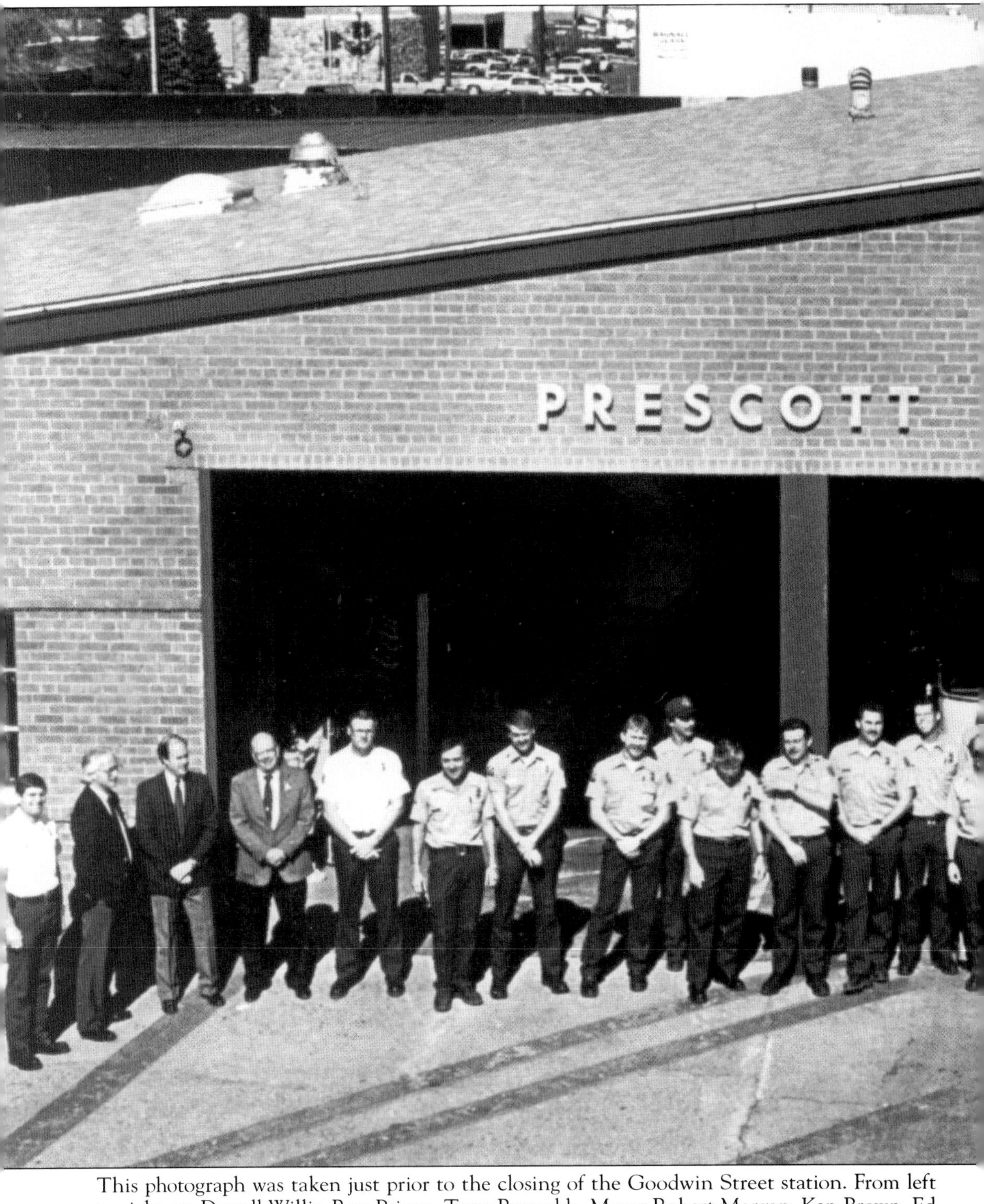

This photograph was taken just prior to the closing of the Goodwin Street station. From left to right are Darrell Willis, Ron Prince, Terry Reynolds, Mayor Robert Morgan, Ken Brown, Ed Hoffman, Dean Hinshaw, Scott MacDermid, Les Roberts, Ted Galde, Jim Barber, Virgil Esmont, Tim Sheehan, Tim McElwee, Kathleen Sinclair, Ron Kayser, Dennis Jeffrey, Duane Steinbrink, Paul

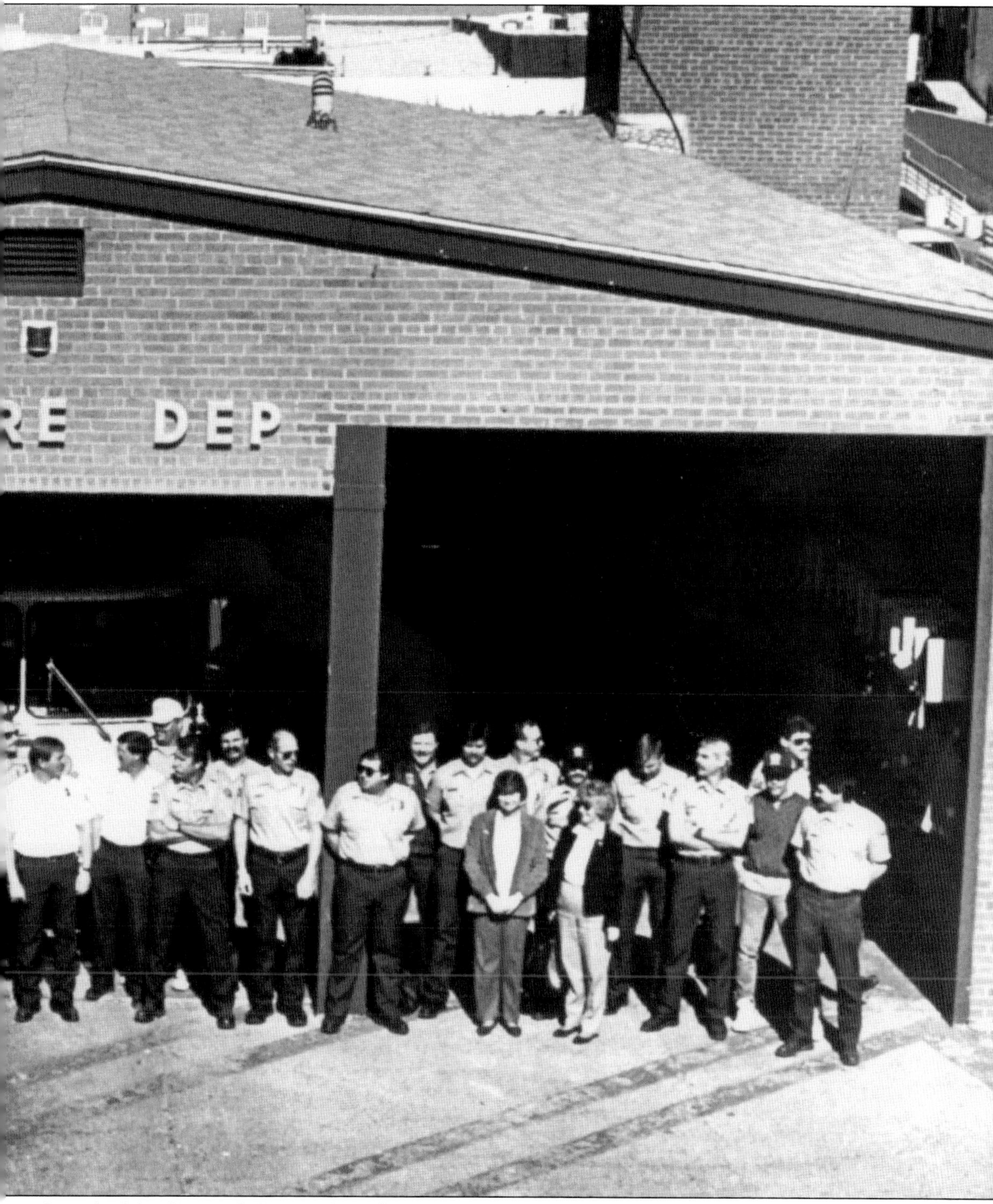

Laipple, John Bostwick, Denny O'Reilly, Greg Fine, Dave Doty, Tom Campbell, Dan Sanders, Jim Mason, Betty Stowell, Perry Hickman, Jeff Moore, Shirley Hamblin, Jeff Knotek, Gary Pederson, Marty Cole, Paul Williams, and Dave Essex.

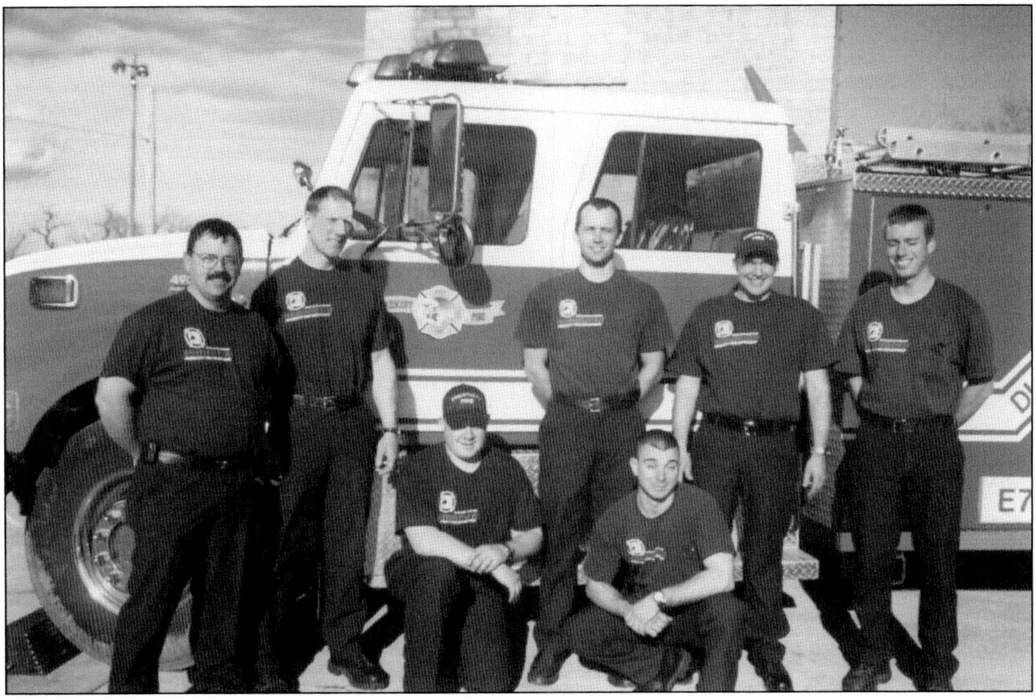

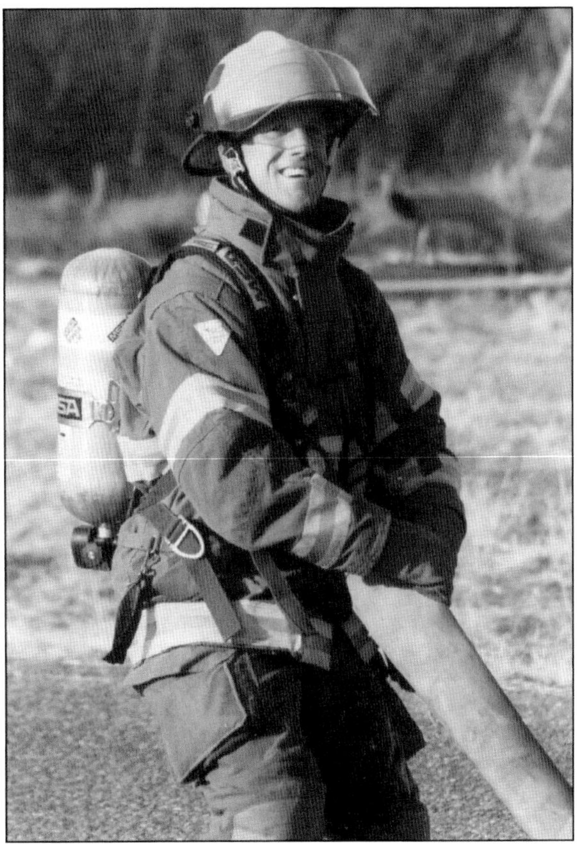

In January 2001, Darren Osborne, Mark Matthews, Troy Steinbrink, Kevin Keith, Dan Bates, Matt Fifield, and Ryan Murphy, pictured here from left to right, were the latest hires. Photographed during probationary training, Murphy and Keith were actually hired in 2000 and gained additional training with the new recruit class. All new hires are run through a minimum of 40 hours of basics in Prescott fire operations before being assigned to an engine.

Mark Matthews was hired in January 2001 at the age of 51, making him the oldest firefighter hired by the department since becoming a paid department. Matthews was a paramedic as well as a member of all the special operations teams on the department. He enjoyed 13 years as a firefighter with the department before retiring in the spring of 2014.

Firefighter Ben Bateman was known for both his biting wit and equally sharp pen. Regularly generating satirical cartoons for publication, he also captured firefighters at their weakest. For a price, Ben would turn a blind eye (and pen) to fire personnel. This caricature is of Jim Mason, drawn with an accentuated abdomen. (Courtesy of Jerry Bateman.)

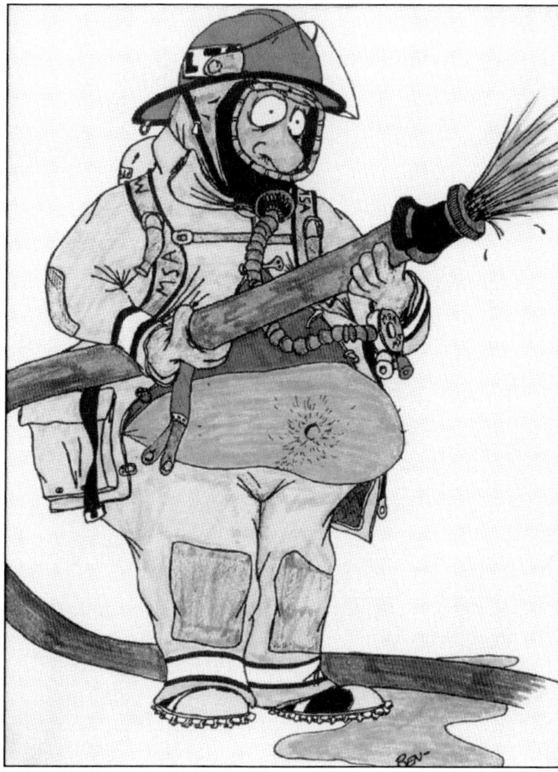

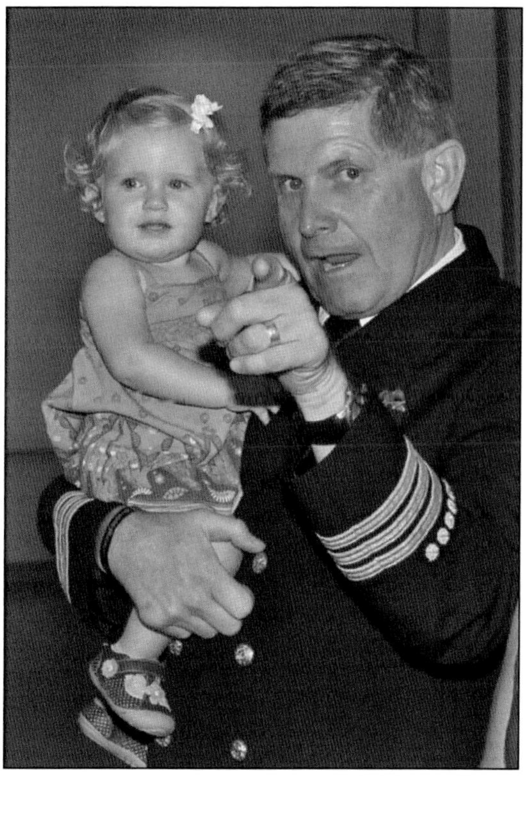

Darrell Willis, seen here with granddaughter Brooke, originally worked for Mesa Fire Department before becoming the deputy chief for Prescott Fire Department. Willis took over as chief from Ron Prince in 1994. He continued as chief until becoming the emergency services director for the city in 2007. After two years as the director he served as a division chief over the wildland division. (Courtesy of Wade Ward.)

Prescott Fire Dep

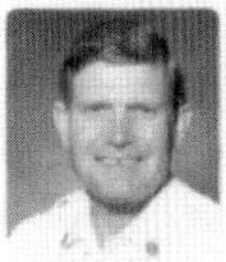

Darrell
Willis
Fire Chief

19

SINCE 1885

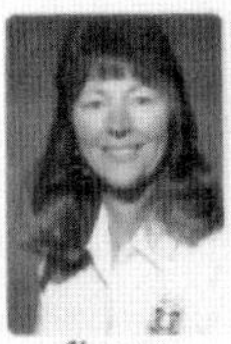

Sharon
Nordyke
Fire Inspector
Prevention

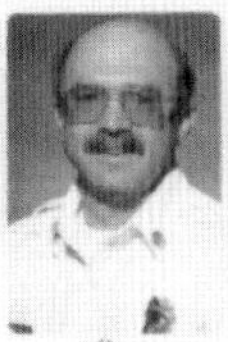

Tim
McElwee
Training/
Communication

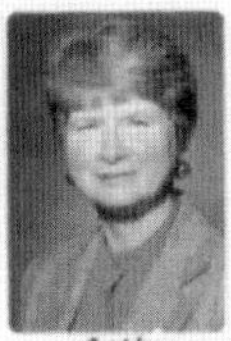

Betty
Stowell
Administrative
Assistant

Battalion Chiefs

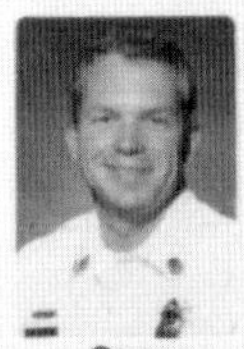

Don
Devendorf

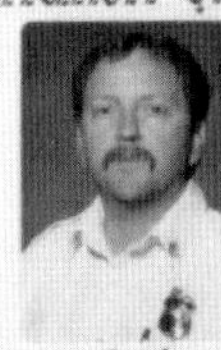

Brad
Malm

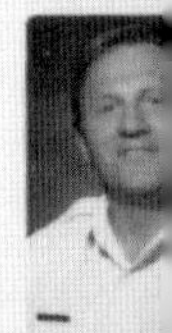

Dua
Steim

Captains

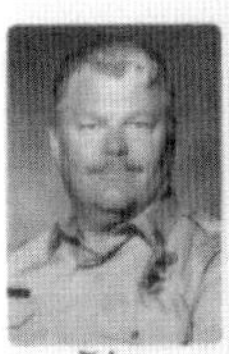

John
Bostwick

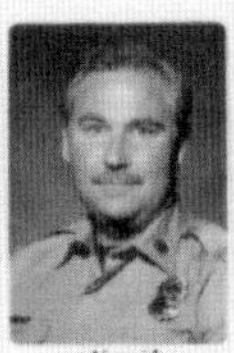

Virgil
Esmont

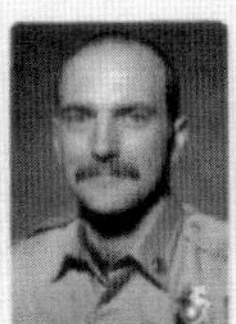

Jay
Fillingim

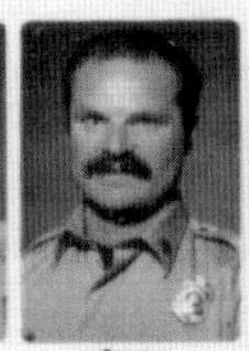

Greg
Fine

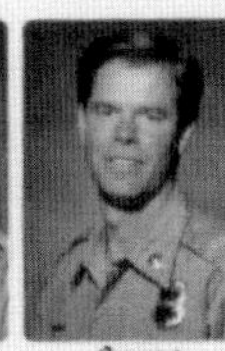

Bruce
Martinez

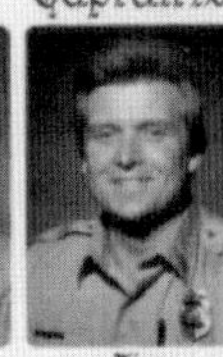

Jim
Mason

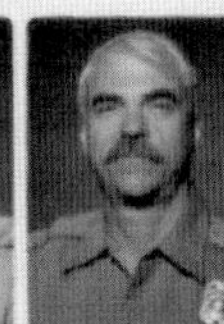

Gary
Pederson

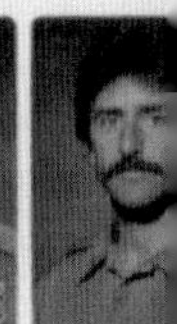

Dav
Scal

Engineers

Not Pictured:
Ted Galde - Fire Marshall
Larry Martin - Captain
Lee Short - Engineer
Steve DelCarlo - Firefighter
Rob Yarbrough - Firefighter

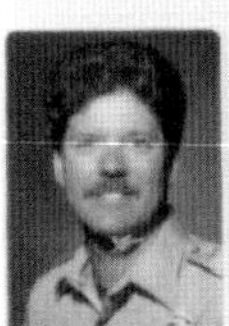

Don
Banning

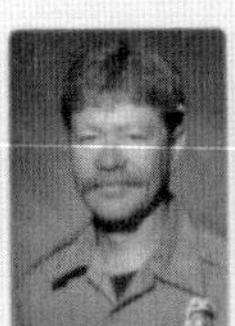

Marty
Cole

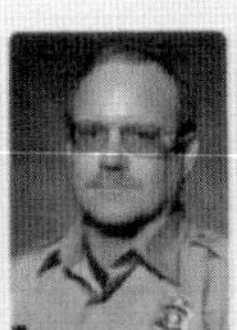

Perry
Hickman

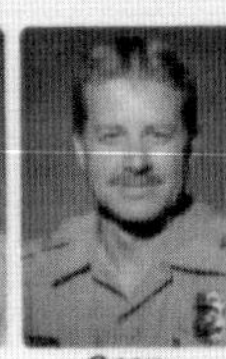

Dean
Hinshaw

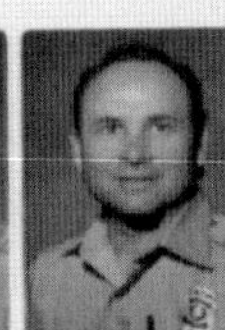

Ron
Kayser

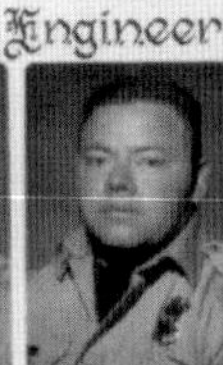

Jim
Kennedy

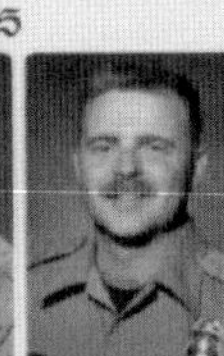

Jeff
Knotek

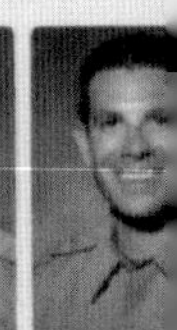

Ralp
Luc

Firefighters

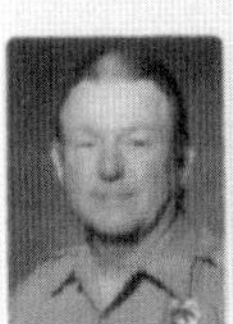

Jim
Barber

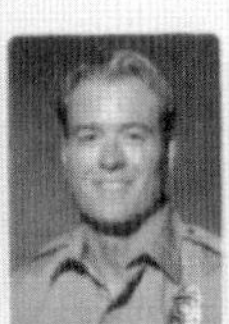

Ben
Bateman

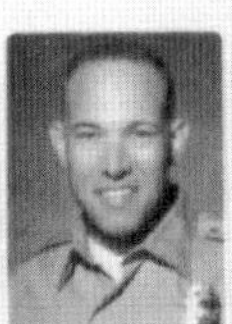

Brad
Colmenares

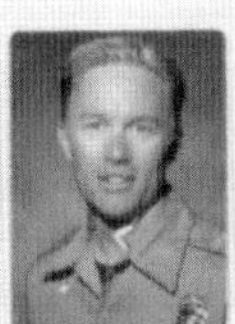

Travis
Epperson

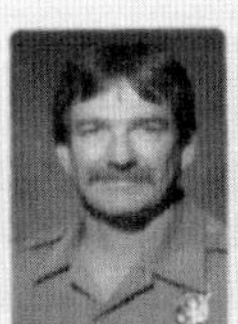

Dave
Essex

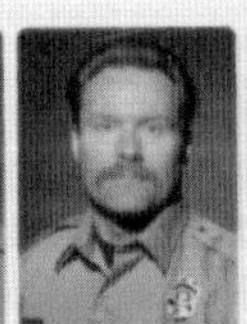

Dennis
Jeffrey

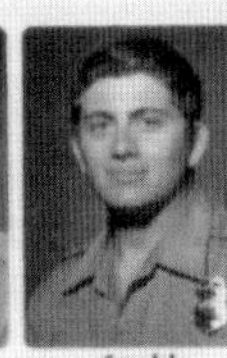

Scott
Luedeman

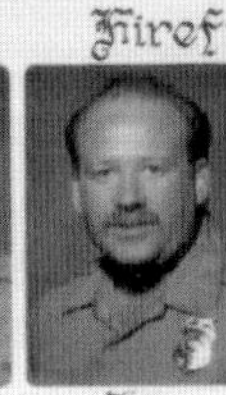

Tim
Miller

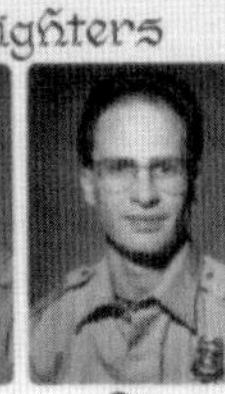

Dan
Morgan

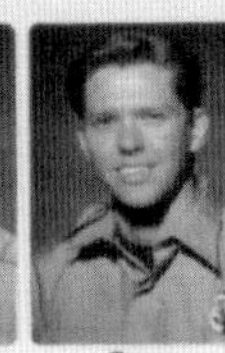

Dave
Peterson

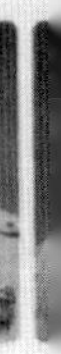

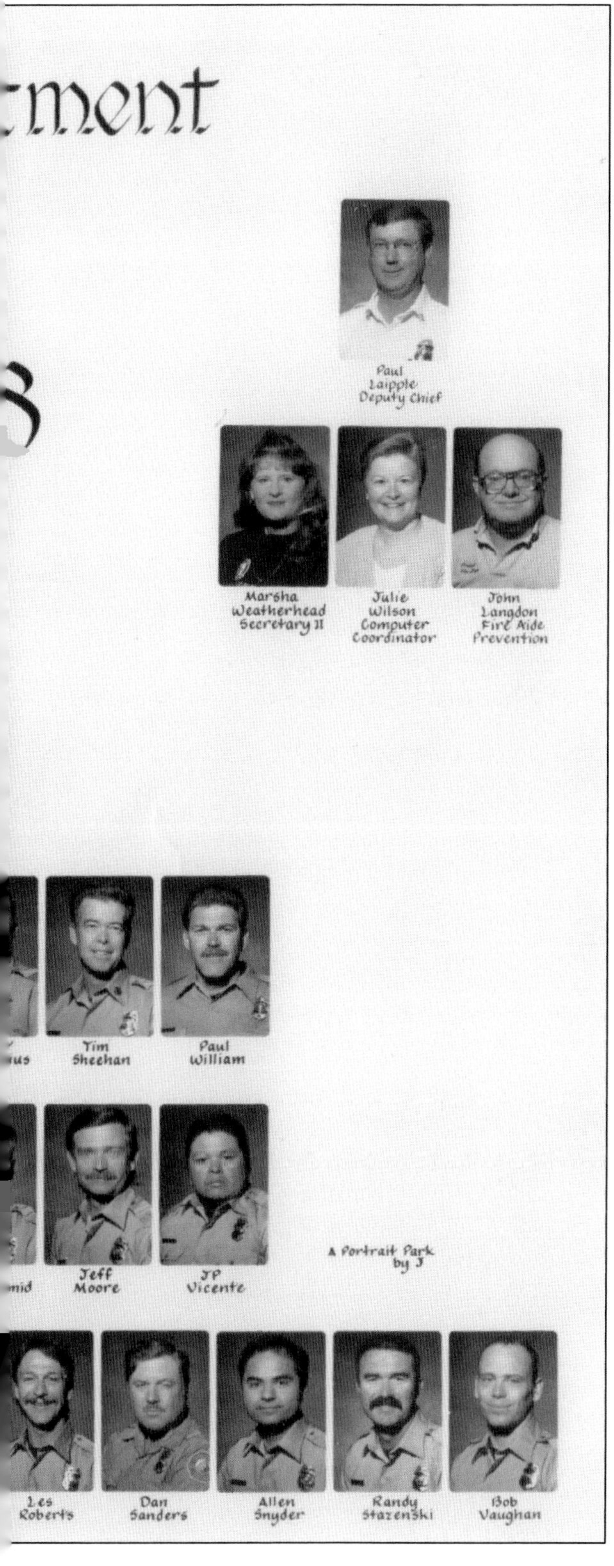

By 1998, the Prescott Fire Department had grown to employ 54 personnel. The population of the community had risen to almost 33,000. Insurance ratings for the area were being influenced by the limited number of stations. In response, Prescott Fire implemented a stopgap measure of creating an Adaptive Response Unit to be run out of the Sixth Street station to increase coverage. Another creative coverage scenario had Prescott and Central Yavapai firefighters sharing engines and stations on opposing shift days.

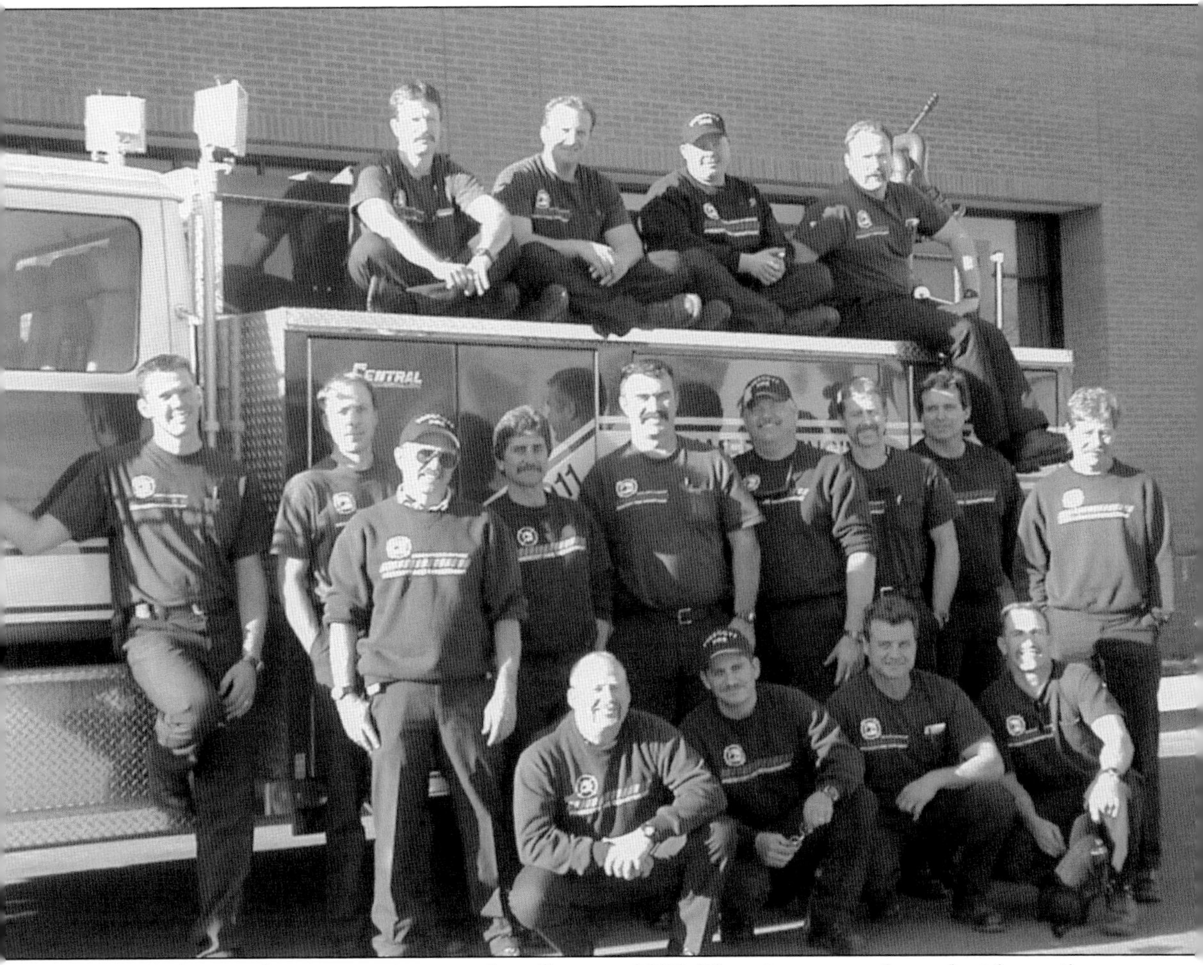

C-shift poses for a group photograph in January 2002. The picture was framed with a toilet seat and given as a parting gift to retiring firefighter Jim Barber, who, as a career firefighter, cleaned the Station 73 toilet for many years. From left to right are (first row) Tim Miller, Scott Luedeman, Larry Martin, and Conrad Jackson; (second row) Dave Peterson, Mark Diedrick, Tim McElwee, Dave Scalia, Randy Stazenski, John Bostwick, Dean Hinshaw, Jim Mason, and Marty Cole. On the engine are Tony Reed, Jason Wallace, Troy Steinbrink, and Dennis Jeffrey. C-shift, as a third platoon, was created in 1975 by Chief Jack Hamilton in order to save the city money as the Fair Labor Standards Act was going to incur extensive overtime costs to the department.

Five

Noteworthy Fires

When firefighters of differing generations sit in a room together, there is always a moment of nostalgia when they compare notes on the large fires they have fought. Prescott Fire Department is no different. With its long history and many annual social and public events, firefighters of previous generations often find themselves mingling with the current generation and swapping stories.

Though none of today's living firefighters can recount the tales of the 1900 fire, it is obvious from newspaper stories in the 1950s that those firefighters were engaged in the same storytelling that occurs now. Some 50 years after the Great Fire, they still could not come to an agreement on the origin of that fire, or even what time the call came out to fight the blaze. Today, retired Prescott firefighters relate stories of major events, including the Foxworth Lumberyard Fire and the Barrows Furniture Store Fire. The current generation is likely to recount stories of the Indian Fire, the Coronado Fire, the Bennett Fire, or the Doce Fire.

The all-hazard response model of the current Prescott Fire Department has increased the variety of calls that the firefighters are called to resolve. Technical rescues have become commonplace, as have hazardous materials events. The bulk of the responses have become medically related, providing a never-ending source of material for storytelling.

There is one story they will all have in common. All members, current and retired, will remember where they were when they learned of the loss of the Granite Mountain Hotshots crew. While not occurring in Prescott proper, but rather some 30 miles south of the city, the Yarnell Hill Fire would have a devastating impact on the Prescott Fire Department, killing 19 of the department's 80 firefighters. With the loss of a quarter of the department, the community was drawn together in support of its historic department, its firefighters, and the surviving families in ways that have been truly humbling.

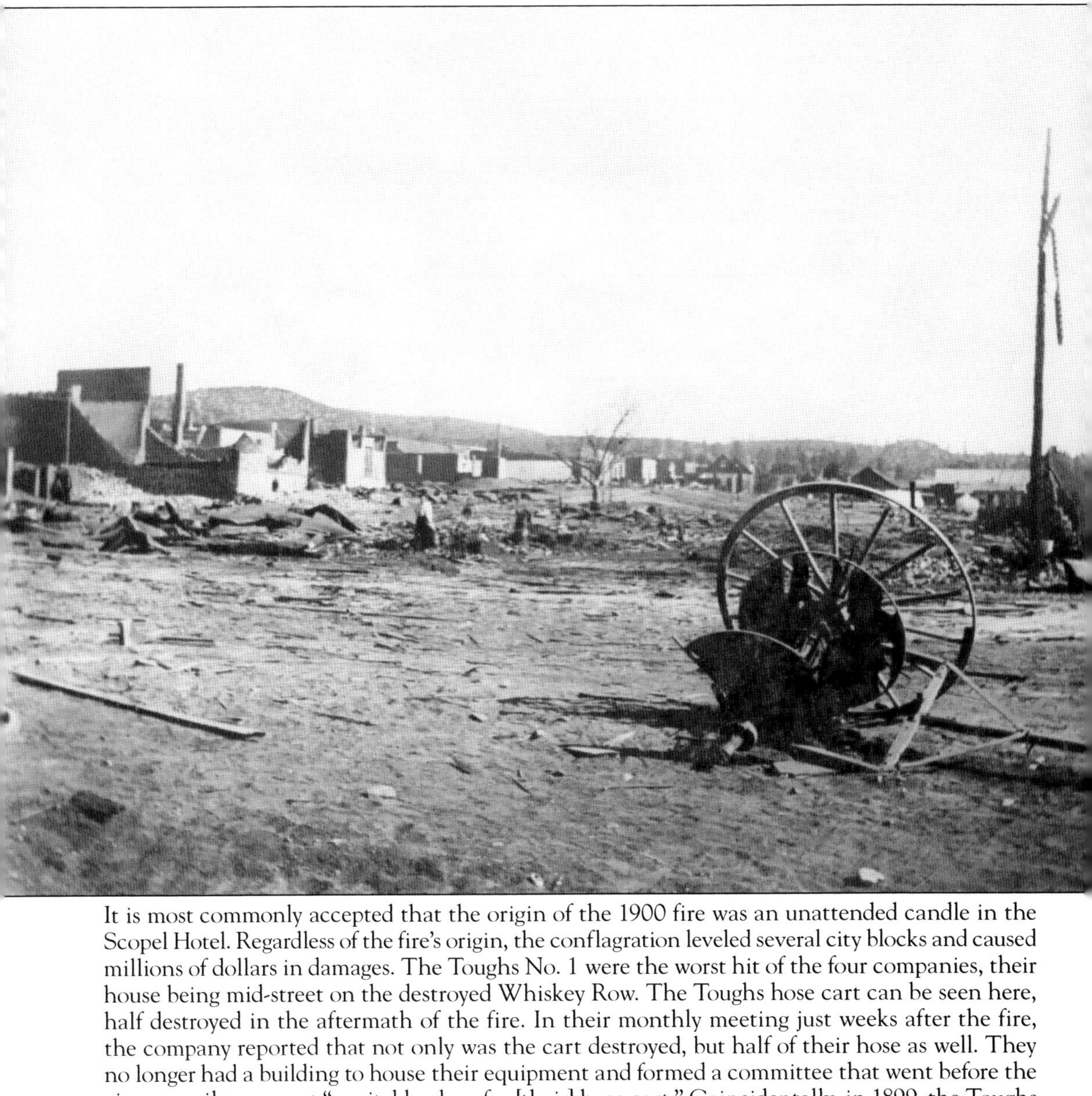

It is most commonly accepted that the origin of the 1900 fire was an unattended candle in the Scopel Hotel. Regardless of the fire's origin, the conflagration leveled several city blocks and caused millions of dollars in damages. The Toughs No. 1 were the worst hit of the four companies, their house being mid-street on the destroyed Whiskey Row. The Toughs hose cart can be seen here, half destroyed in the aftermath of the fire. In their monthly meeting just weeks after the fire, the company reported that not only was the cart destroyed, but half of their hose as well. They no longer had a building to house their equipment and formed a committee that went before the city council to request "a suitable place for [their] hose cart." Coincidentally, in 1899, the Toughs had looked into moving their equipment to the grounds behind the G.C. Ruffner barn, the very same location the whole department would move to in 1906 on the southeast corner of Cortez and Goodwin Streets. (Courtesy of Sharlot Hall Museum, #f2108pe.)

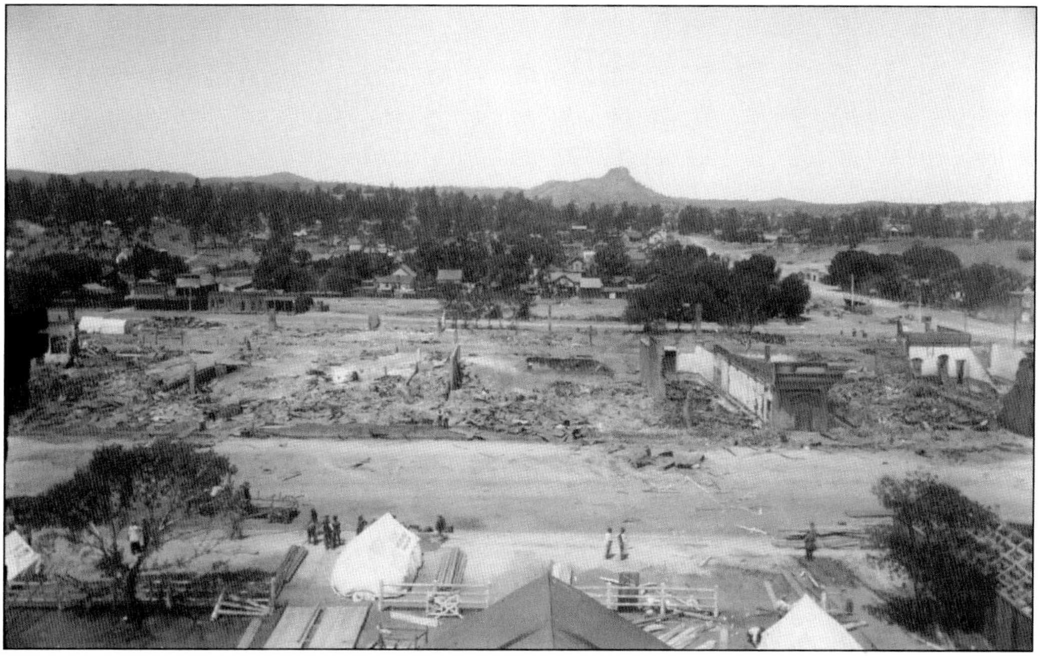

This photograph, taken from the courthouse balcony, shows Whiskey Row after the fire. Two men can be seen standing on what was the sidewalk on the far side of the street. They are standing where the Toughs hose cart had been staged, in a small area between a tailor and a barbershop. (Courtesy of Sharlot Hall Museum, #f2110pa.)

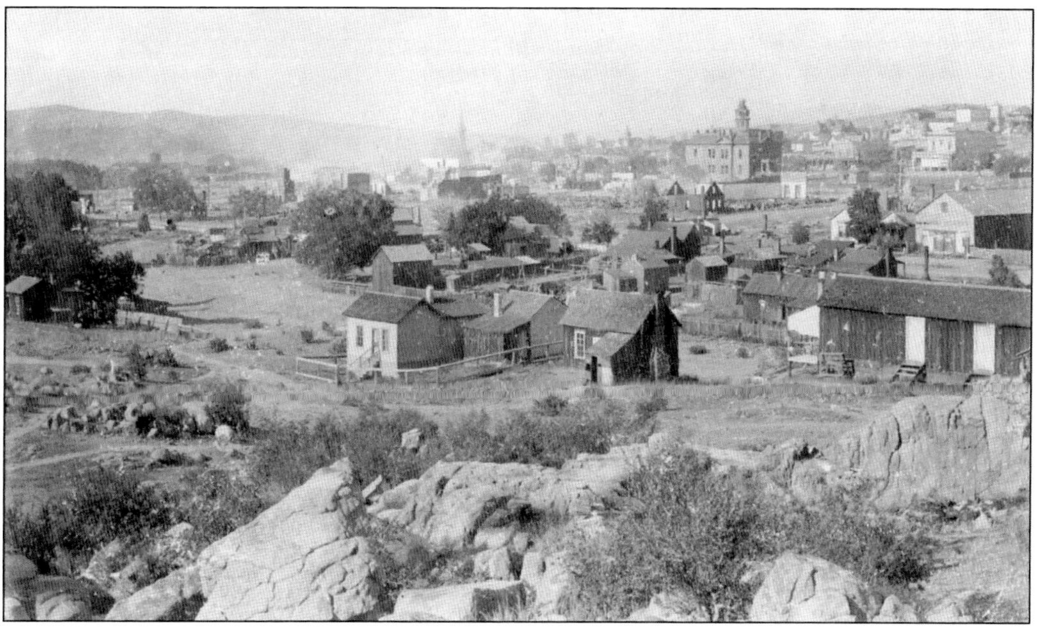

The whole center of Prescott and its business district was gutted by the fire. The majority of the businesses were uninsured. Estimates are that over $2 million in damages occurred. This photograph was taken from the hill where the Arizona Pioneers Home currently sits, looking to the northeast. The courthouse is seen clearly just right of center. (Courtesy of Sharlot Hall Museum, #97.109-24.)

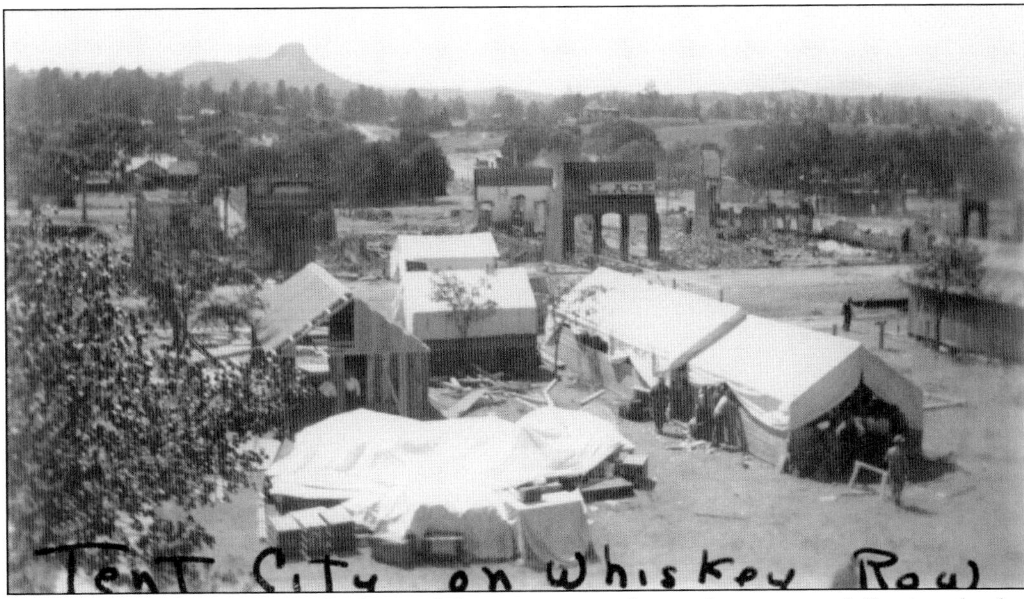

A veritable tent city rose on the lawn of the county courthouse the morning following the fire, with some 50 temporary structures erected. In subsequent days, the structures became more solid as small buildings were erected so businesses and bars could maintain operations until the brick-and-mortar buildings of Whiskey Row could be rebuilt. The peak of the courthouse can be seen rising above the new temporary structures. In the coming years, a new water system and reservoir became a priority so residents would never again experience dry wells in their time of need. (Both, courtesy of Sharlot Hall Museum: above, #f2103pd; below, #f2103ph.)

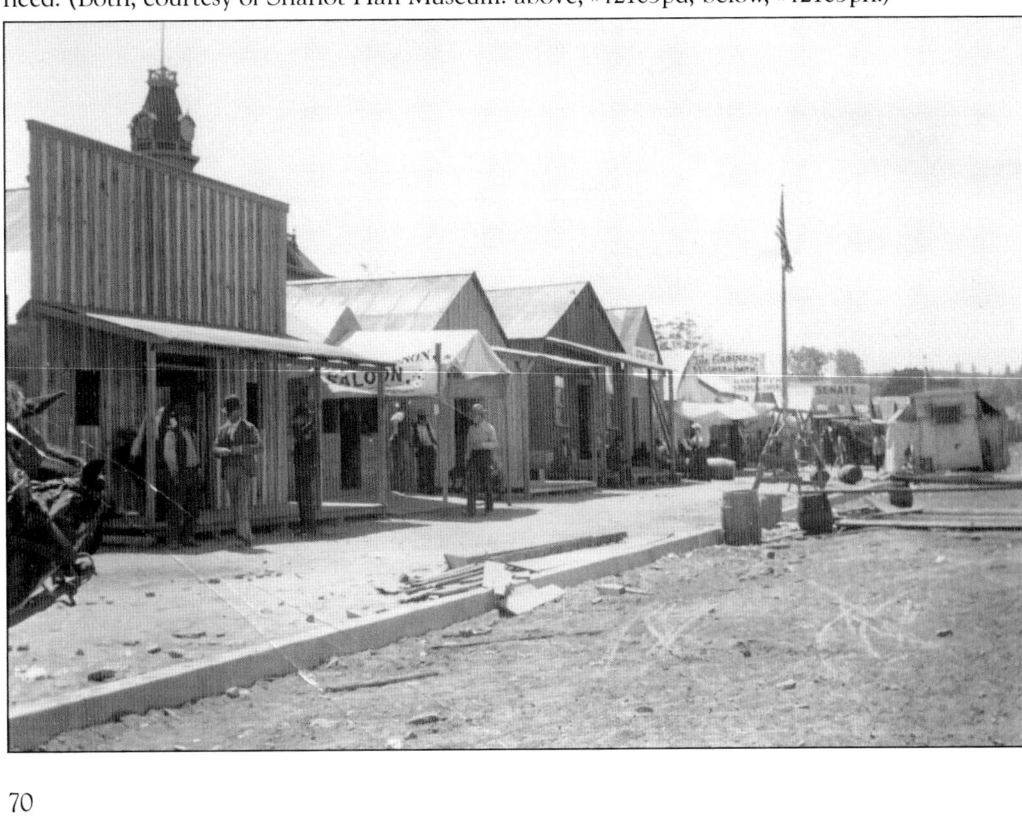

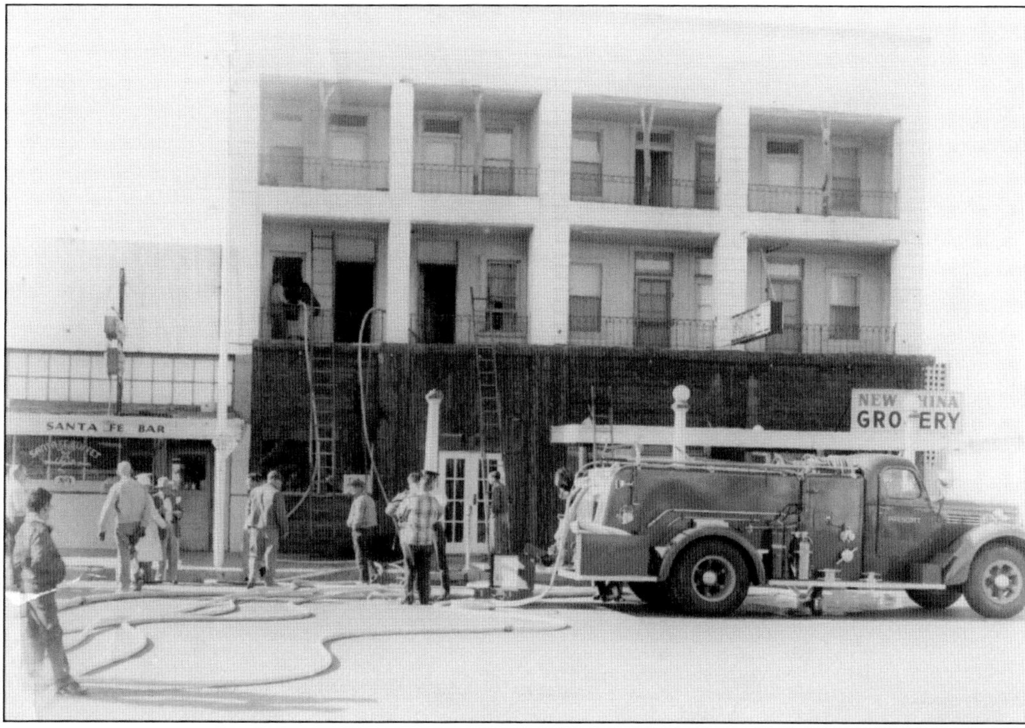

Above center and at right, on Christmas day in 1954 in front of the former Harvey Hotel on North Cortez, George Ravetto puts a self-contained breathing apparatus on Ernie Brown. The building is hardly recognizable today, as the front balconies have since been covered over and the upper stories are now used for storage. The Scott brand SCBA pictured is one of the first used by the department and was of the demand-air variety with a belt-mounted regulator. The department at the time had two 15-minute units and six one-hour varieties. The devices were used only sporadically, unlike today when the presence of toxic chemicals in smoke is recognized as highly hazardous and masks are worn throughout operations.

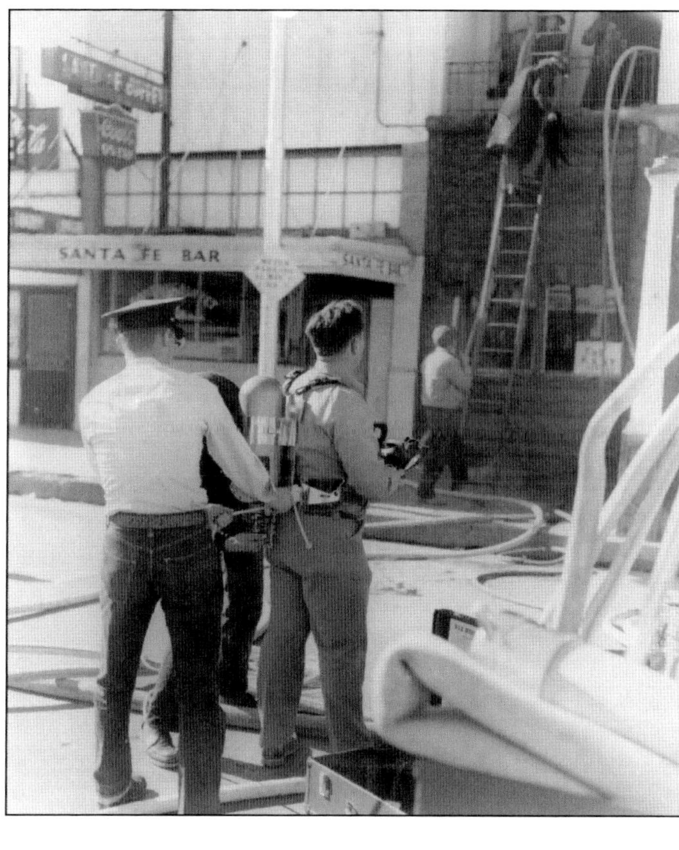

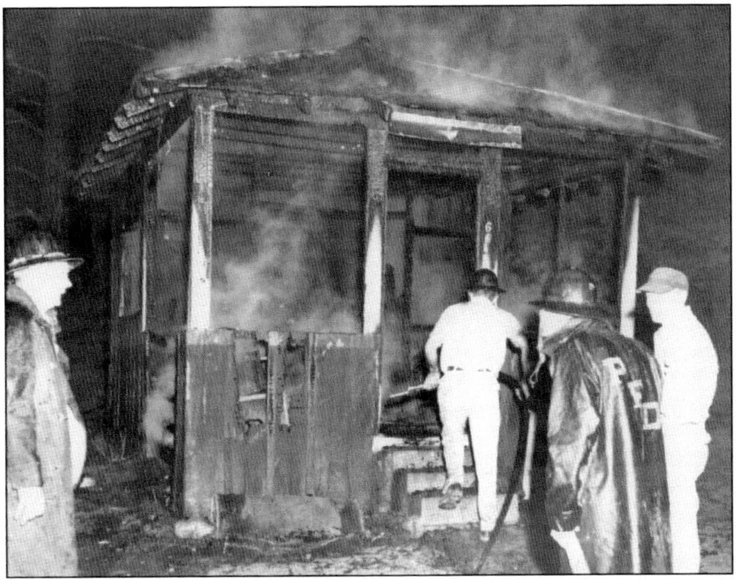

On June 16, 1957, firefighters Mac Reininger, Bob Eddings, Bill Robertson, and John Bouton fight a fire in a cabin at the Arizona Church Conference Center, which would later be named Chapel Rock. Though only four are seen in the picture, this small fire had a respectable turnout, with 26 of the 35 volunteers showing up at the incident.

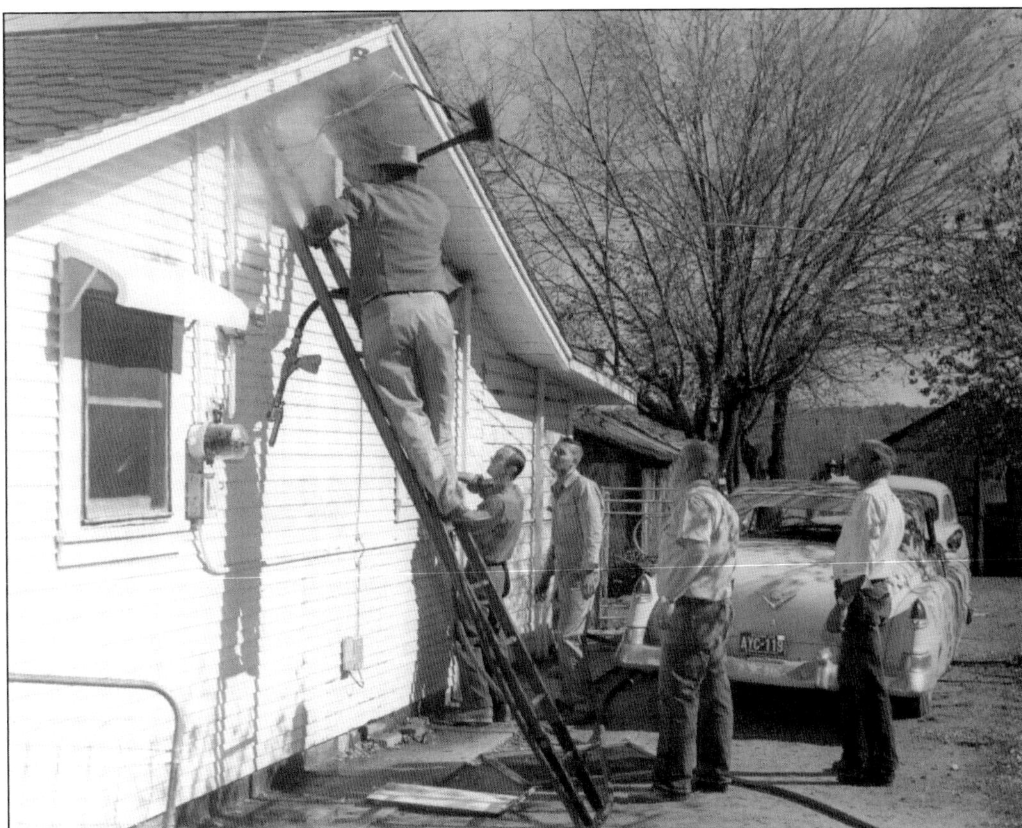

The temptations faced by children never seem to change. On November 23, 1960, this home fire at 815 Lincoln Avenue was started by a three-year-old playing with a lighter. These firefighters are attempting to gain access through the eave as the fire spread through the attic. Ultimately, the fire spread through the entire home.

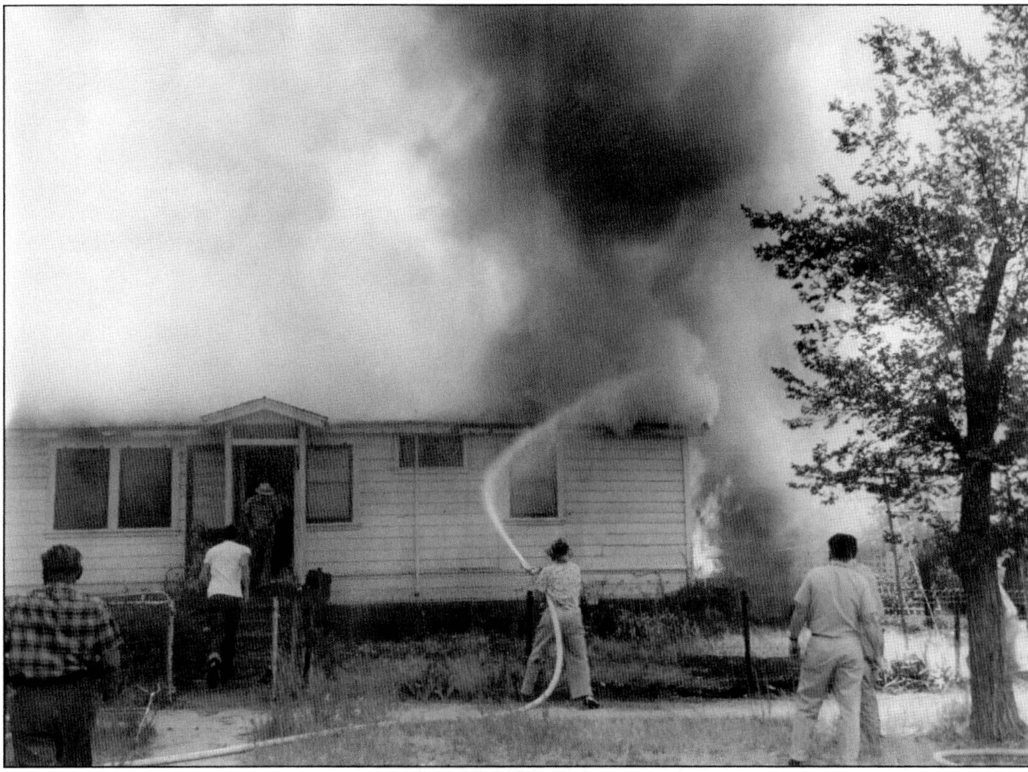

These two photographs from the 1950s demonstrate the need for further professionalism and safety within the department. Insurance underwriters of the time recognized the need for transition as well. The city was faced with a change in its insurance rating due to its lack of a paid, fulltime chief and firefighters as well as antiquated equipment and inadequate water supplies. The city barely maintained its "five" insurance rating by two points in 1957. The city actively sought to improve its shortcomings with new engines and stations. The insurance rating dropped from "five" to "four" in 1965, saving the residents and business owners an estimated $140,000–$180,000 in their insurance premiums.

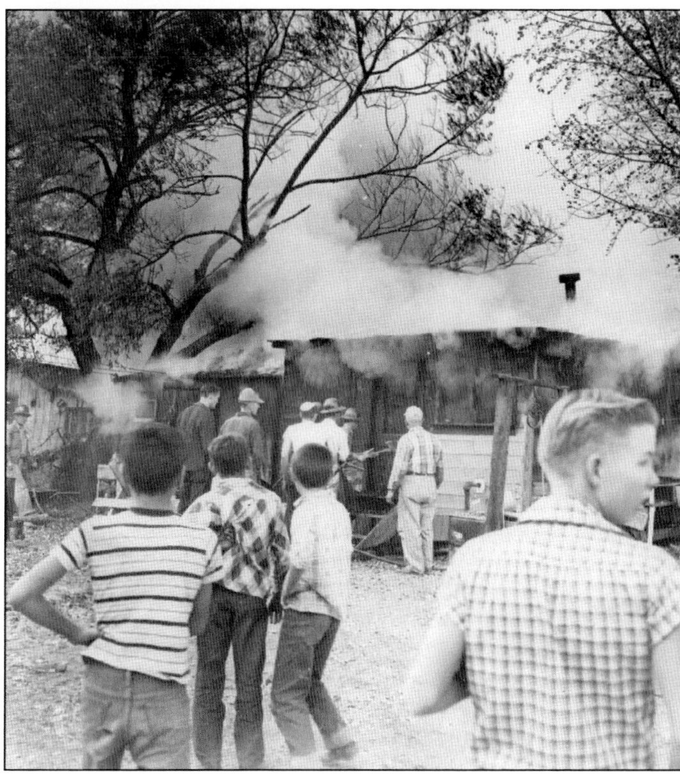

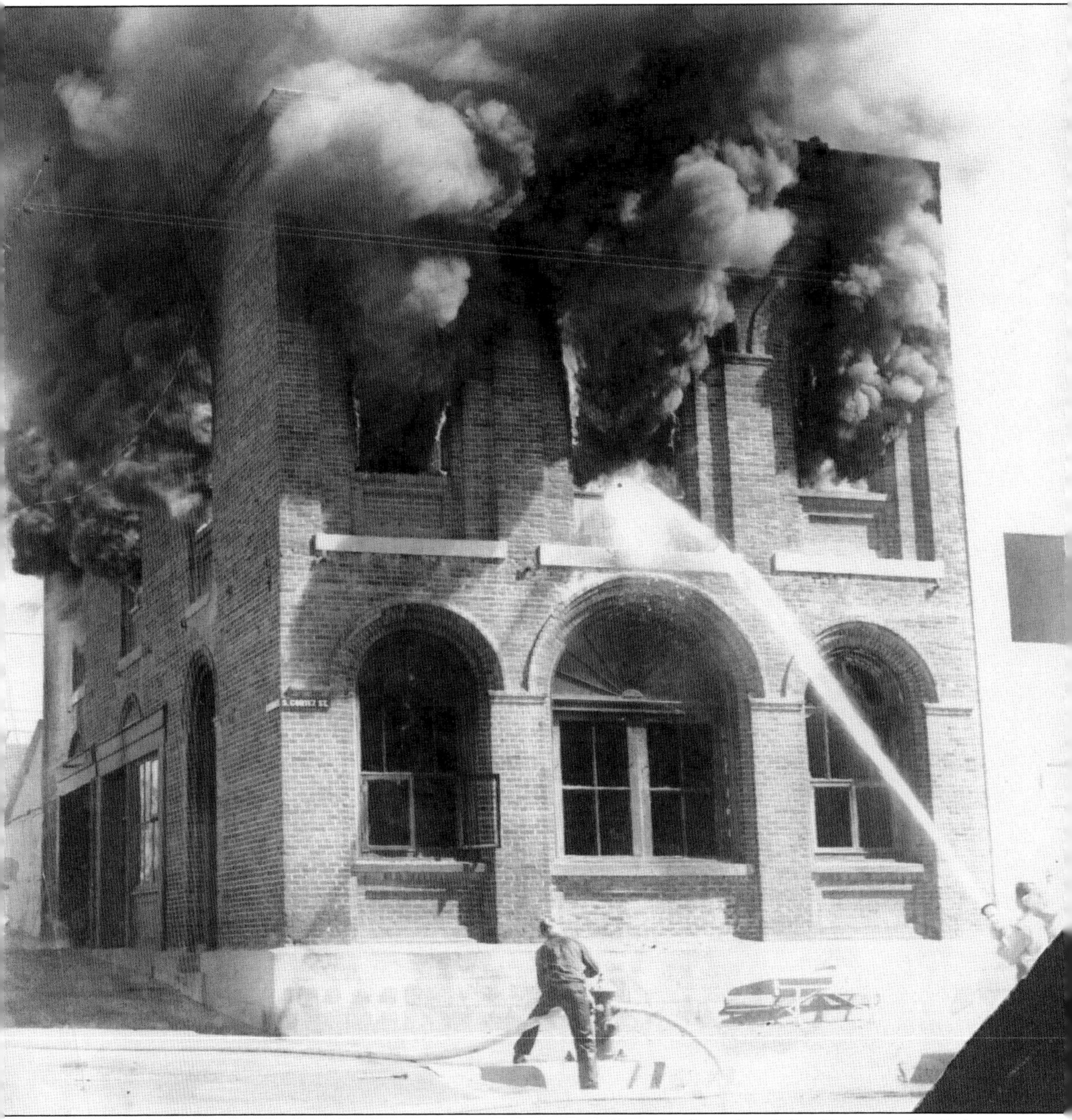

Howey Hall and the adjacent firehouse met their end at the hands of the Prescott firefighters, intentionally. Once the new fire station was built on Goodwin Street, the city retired the structure with the intention of building a new city hall on the site. The lowest bid for the demolition of the building was $700, more than the council was willing to pay. Chief Custer proposed burning the building as training in an effort to teach his firefighters to live with fire and "respect it but not fear it." Custer reported to the council that a third of his firefighters had no experience against an actual fire. Firefighters repeatedly lit fires in the building over the course of several weeks in December 1958 and into early 1959. The mayor and council were present one evening to witness how little water was required to extinguish an incipient fire, a fact that reinforced the importance of timely responses by firefighters.

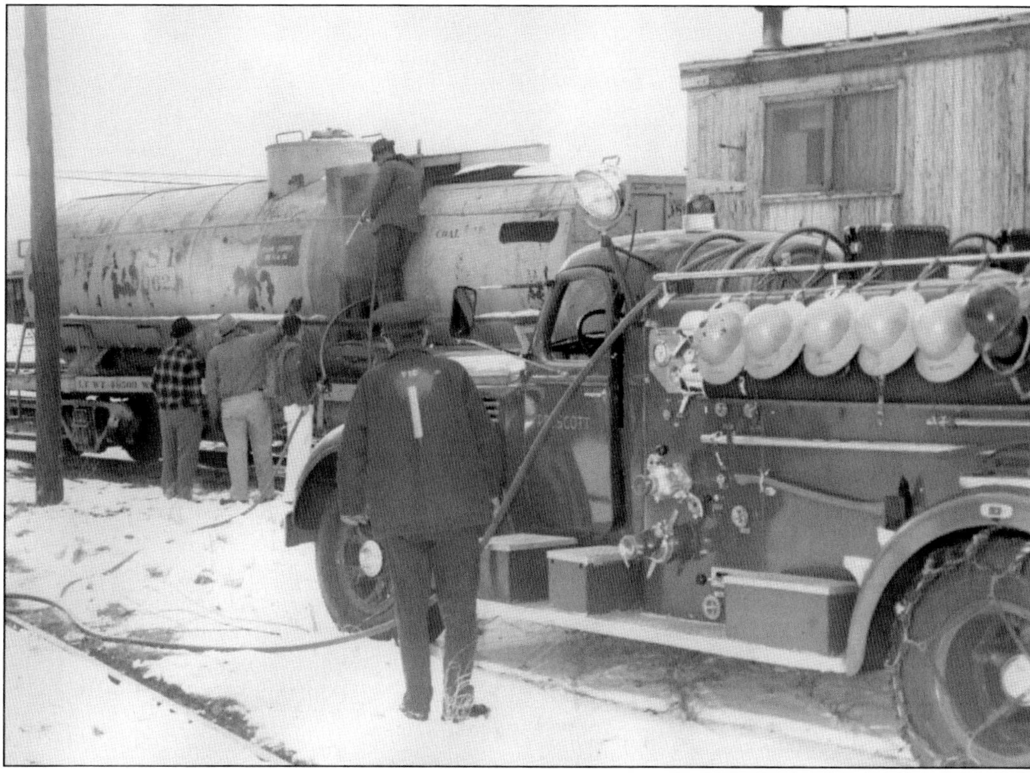

February 1962 saw the firefighters responding to the Santa Fe Railroad Depot for a fire in the coal and water car. A flare being used to thaw out a frozen pump ignited some sawdust packing material. Helmets hanging from the side of the engine were intended to be available to firefighters responding directly to the scene.

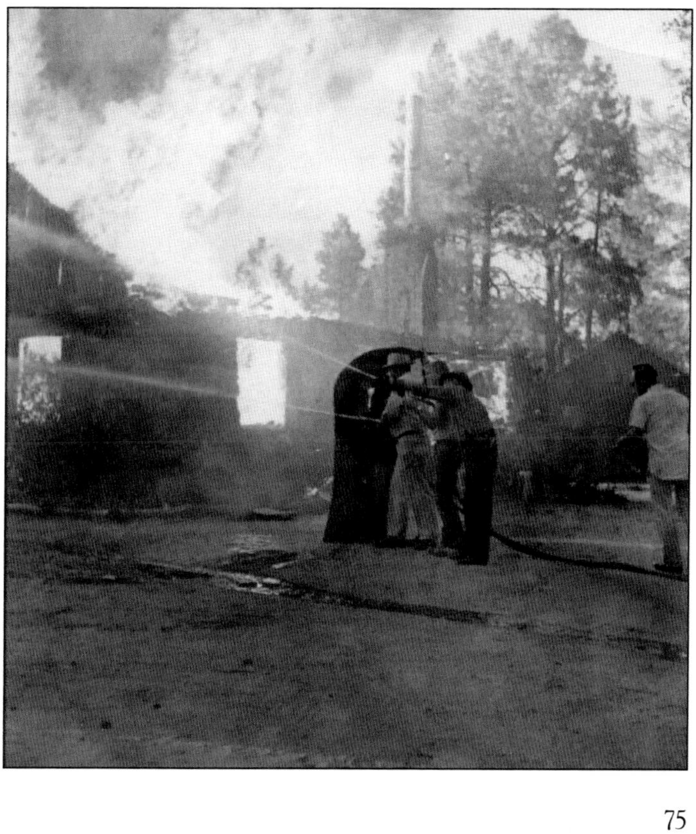

Firefighters in this 1962 photograph hold up a sheet of metal to reflect the intense heat as they fight to prevent the spread of this house fire in the Seventh-Day Adventist Camp. Started by a gas furnace, none of the family of six was injured in the fire.

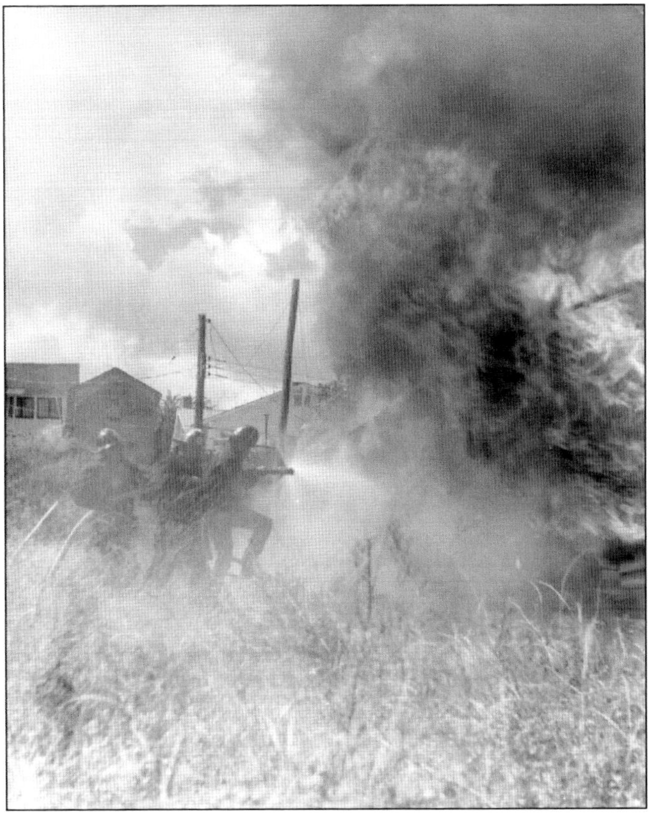

After the precedent of intentionally burning Howey Hall for training was established, it became a common practice. Firefighters would assist homeowners who wanted to remove a building from their property by burning it down for them and using the opportunity to practice their firefighting skills in the process. The practice became so popular that Prescott Fire Department began charging for the service, $1 per square foot. It would later be discontinued as a training tool in the early 2000s due to safety concerns. Four Prescott firefighters were sent to the hospital in 1975 when a back draft occurred during this type of training.

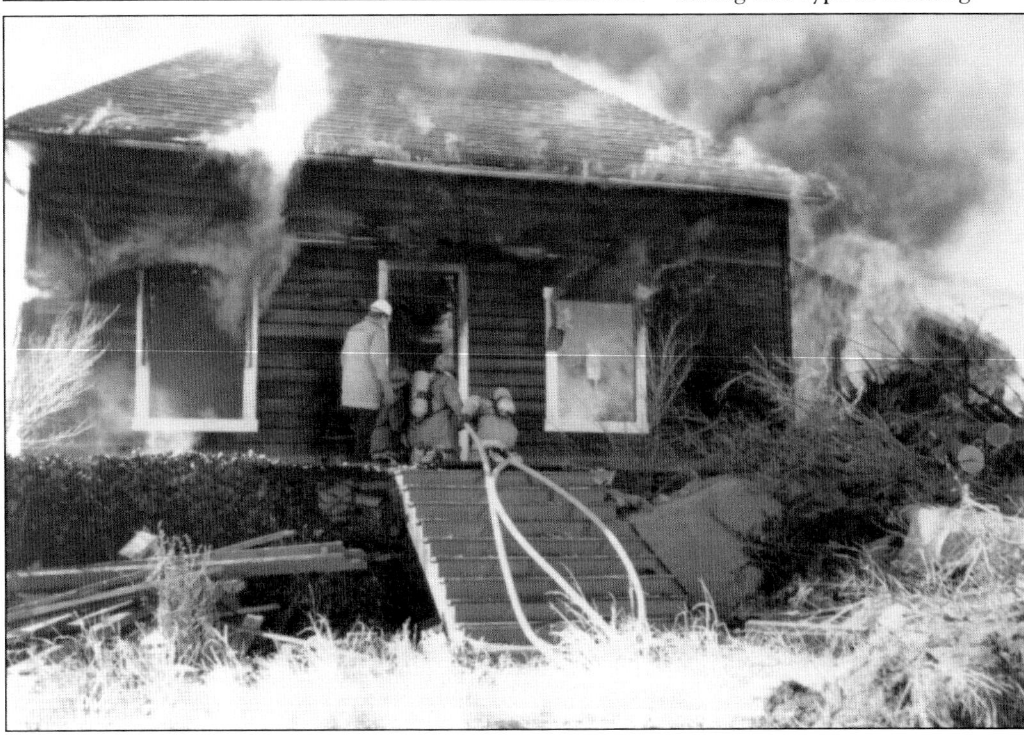

Willy Davis, Bill Wagner, Art Crawford, and Tom Campbell stand on the 1931 Seagrave truck with Rod Owen looking up from the ground. This shot was taken on the corner of South Cortez and Union Streets and shows the marquee for Studio Theatre in the background, creating a composition no photographer could resist. Firefighters responded to the Barrows Furniture Store Fire around 1:00 a.m. on May 12, 1972, and fought the fire through the night. In the image below, Dennis Kasey and Lloyd Laub sit amongst a spaghetti pile of hose lines while managing their nozzles. (Both, courtesy of Wendell Wilson.)

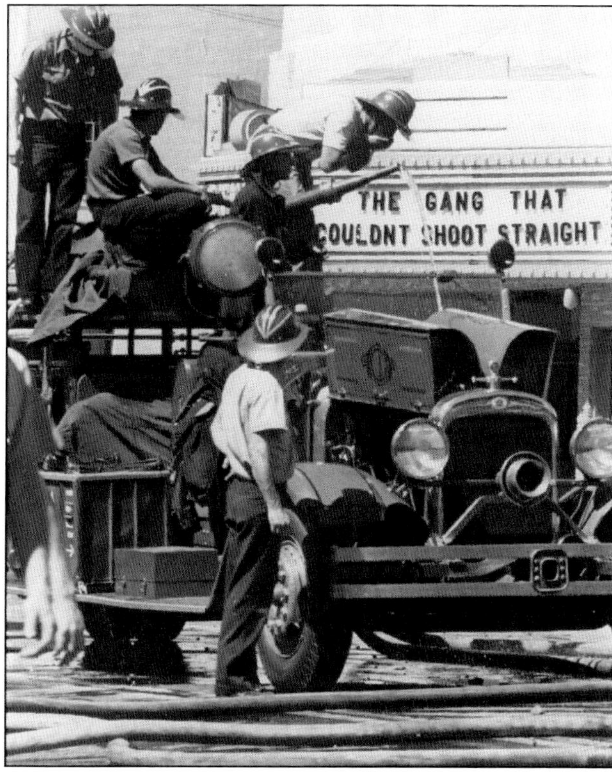

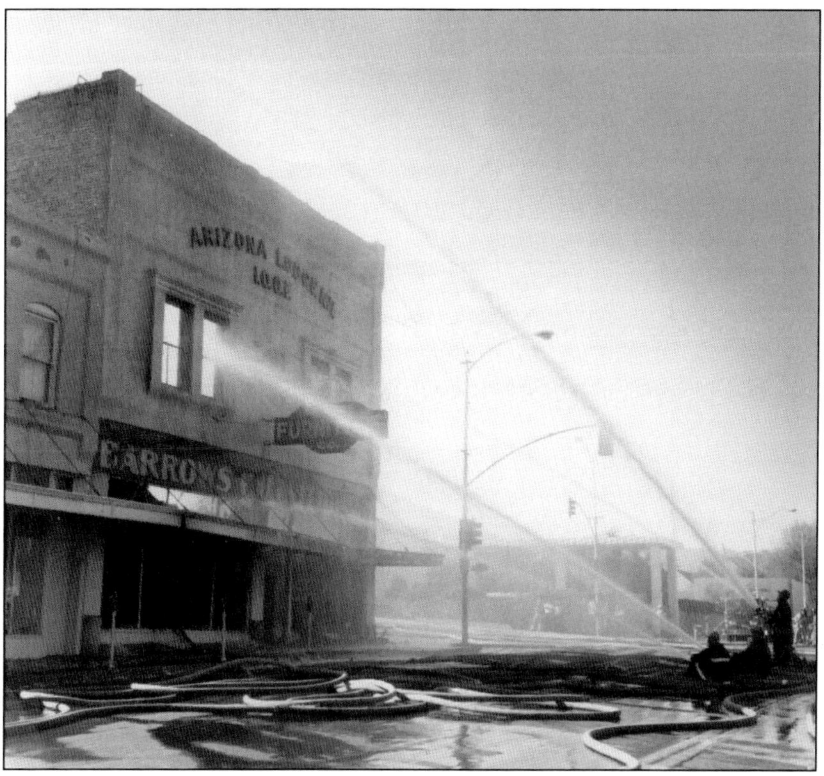

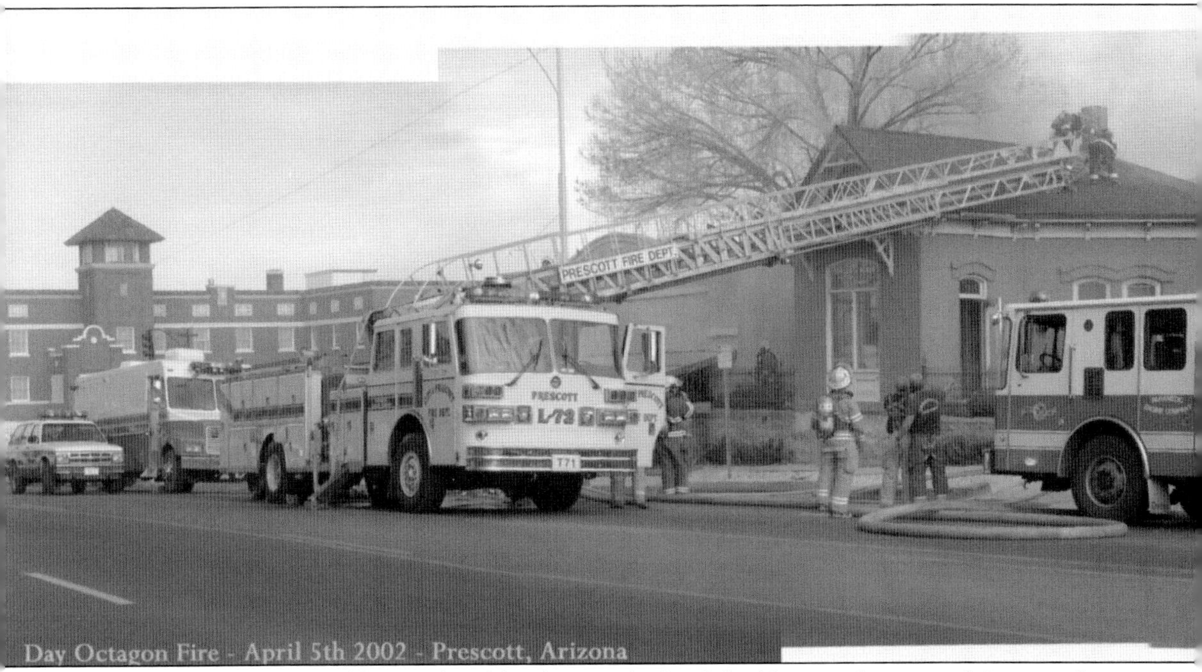

The Day Octagon building is the oldest known firebrick building in the state of Arizona. It was built in 1877 for Dr. Warren Day and served as the first surgical hospital in Arizona. Firefighters responded to a fire in the building on April 5, 2002. The fire occurred during the morning shift change, which boosted the number of firefighters available to fight the blaze. The fire started

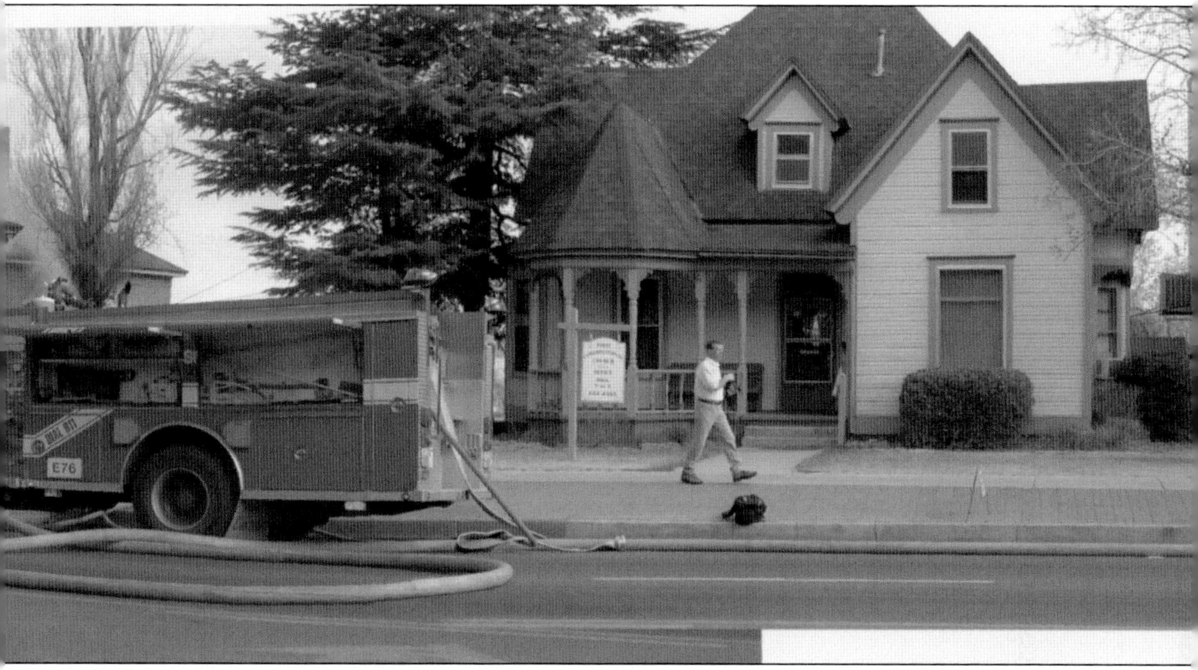

in the basement and extended into the attic before it was stopped. This fire was also one of the first where the department's new thermal imaging camera was used during suppression activities. (Courtesy of Jeremy Cox.)

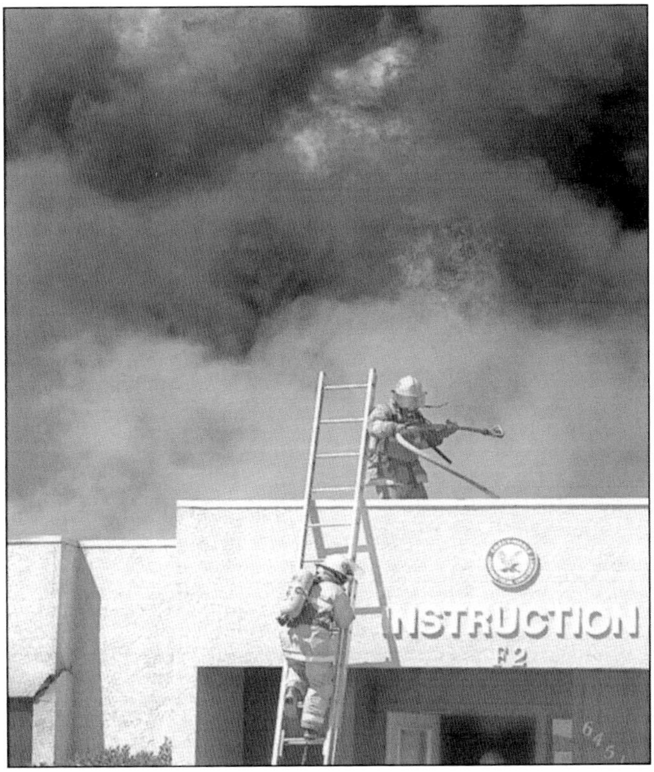

Dave Essex pulls a hose line as he beats a hasty retreat off the roof of this Embry Riddle Aeronautical University training building. He and Travis Epperson had been attempting a trench cut to stop the progress of the fire. The roof proceeded to vent and forced their evacuation. Don Devendorf stands on the ladder while observing Essex's progress.

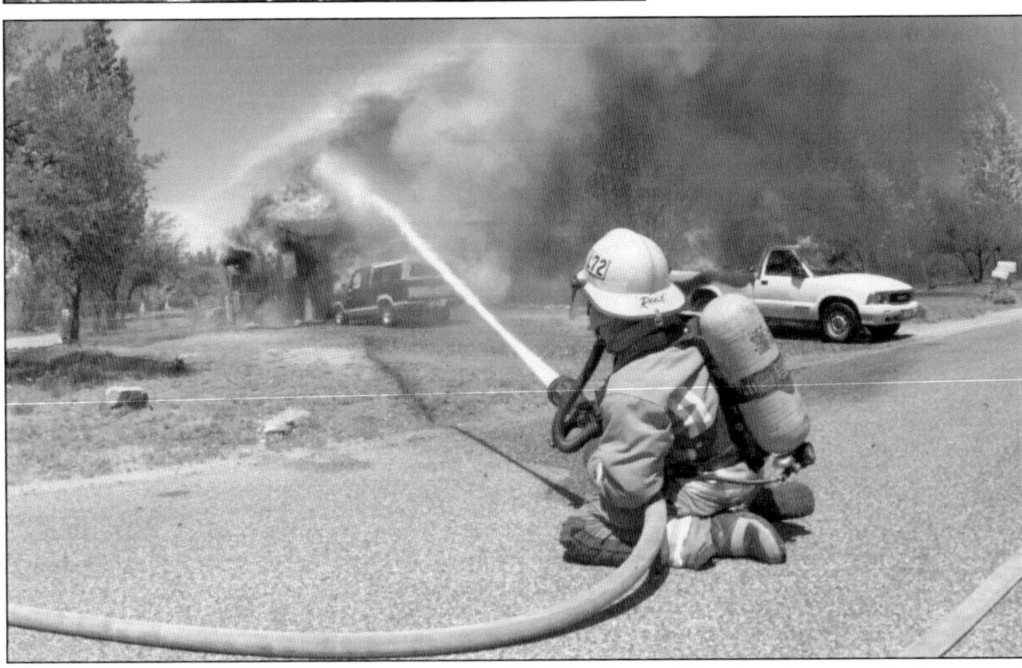

Tony Reed sprays water onto this home in the Ho Kay Gan subdivision from a safe distance. A large stock of ammunition and reloading equipment in the home made firefighting difficult and made the area sound like a war zone. The 1,400-square-foot home and two trucks were destroyed as the neighborhood residents sheltered in place. (Courtesy of Les Stukenberg.)

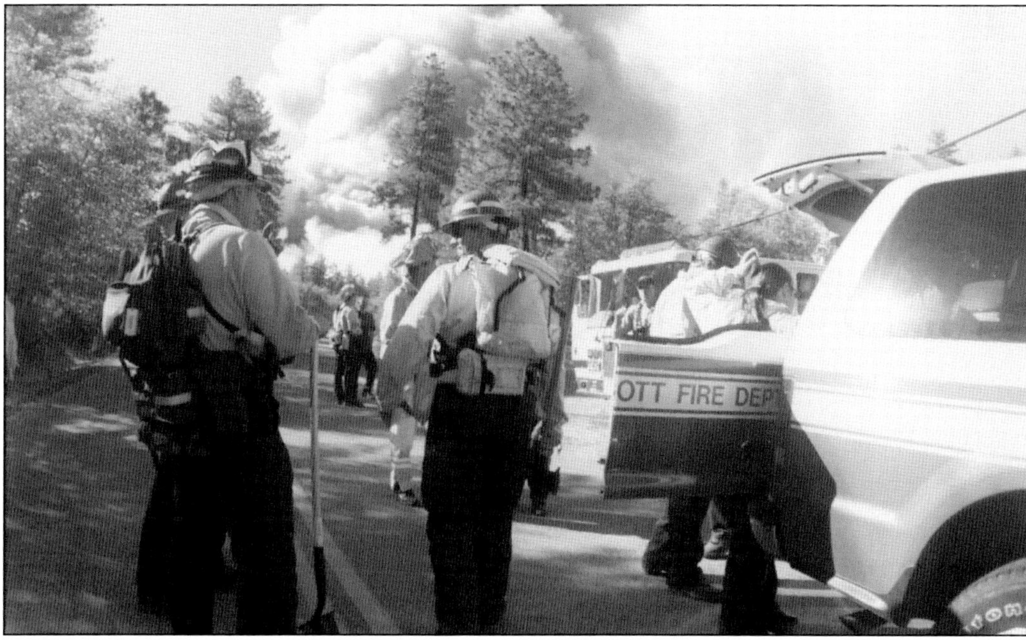

Starting south of Prescott near Indian Creek campground on May 15, 2002, the Indian Fire ultimately charred some 1,000 acres, forced over a thousand people to evacuate their homes, and destroyed seven homes and outbuildings before being halted. The blaze drew attention to Prescott's ongoing wildland-urban interface issues, the importance of having adequately trained wildland firefighters on the department, and the value of the region's cooperative agreements among its firefighting agencies. (Courtesy of Les Stukenberg.)

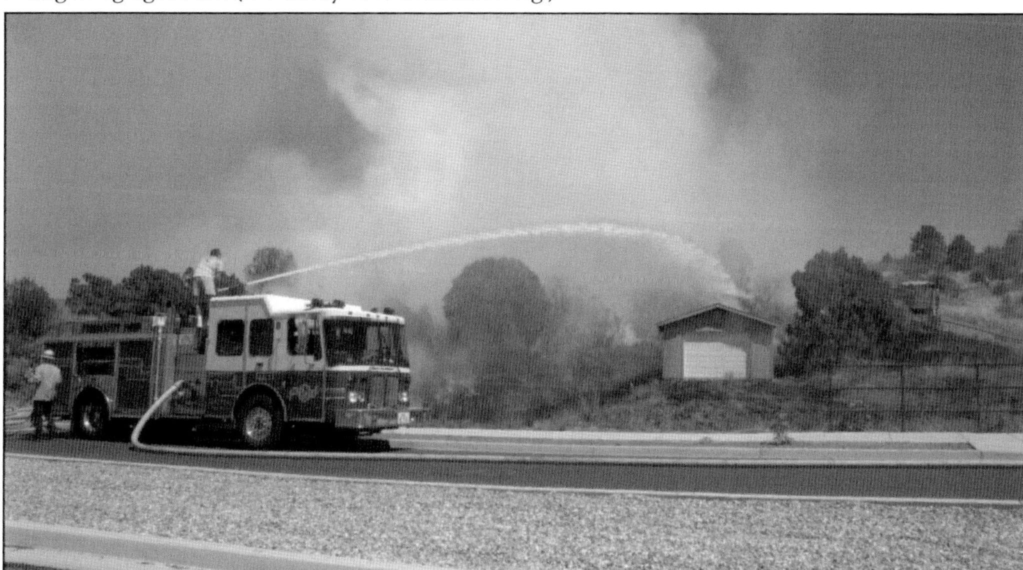

This wildland fire along Rosser Street threatened the home of state representative Lucy Mason. Firefighters stopped the blaze without loss of structures. Seen here, an off-duty firefighter uses the deck gun and an ample supply of water from a hydrant to keep the fire out of a shed. Crew 7 was dispatched to the fire to construct hand line around the fire, demonstrating the need for a wildland division.

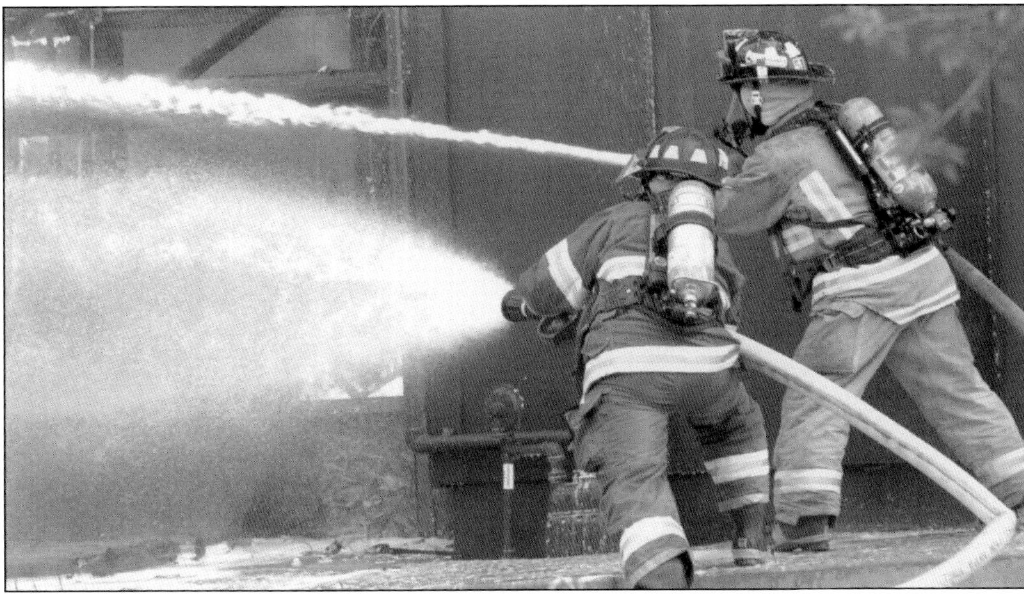

The vacant Roadhouse 69 restaurant caught fire on May 10, 2005, when a transient attempting to commit suicide lit the building on fire. Firefighters from Prescott Fire Department, Central Yavapai Fire District, and the US Forest Service were able to contain the fire to the building and stop it from spreading to the adjacent Granite Creek Park. The city would later purchase the property for a new fire station, but those plans fell through when the economy failed. (Courtesy of Les Stukenberg.)

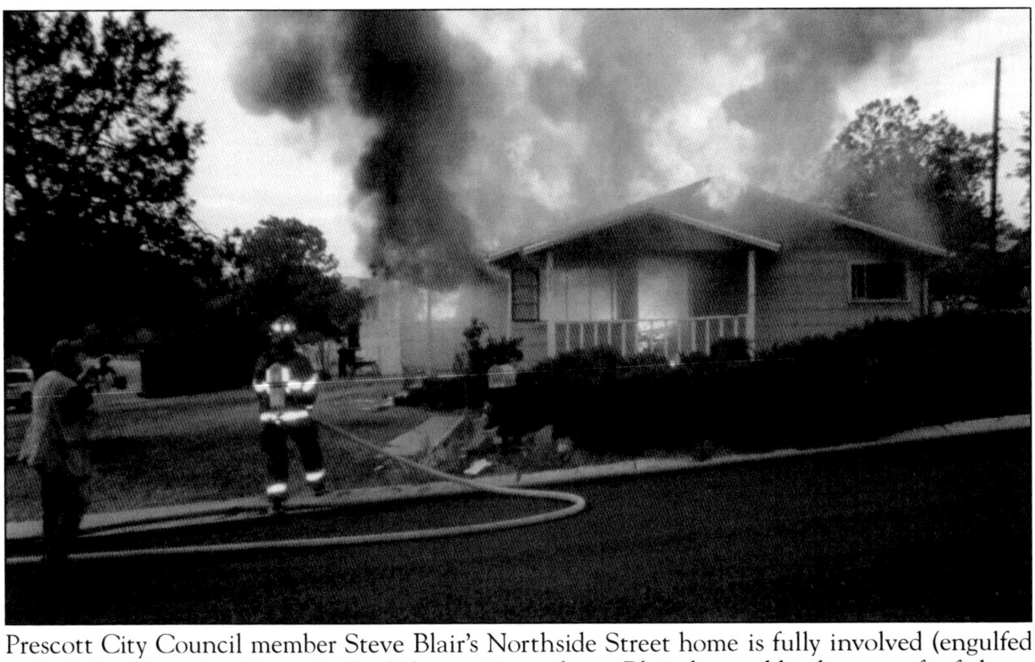

Prescott City Council member Steve Blair's Northside Street home is fully involved (engulfed in fire), but intentionally set by firefighters. Councilman Blair donated his home to firefighters for training in the weeks prior to its scheduled demolition. The building was used primarily for practicing roof ventilation cuts as well as search and rescue techniques prior to being burned to the ground.

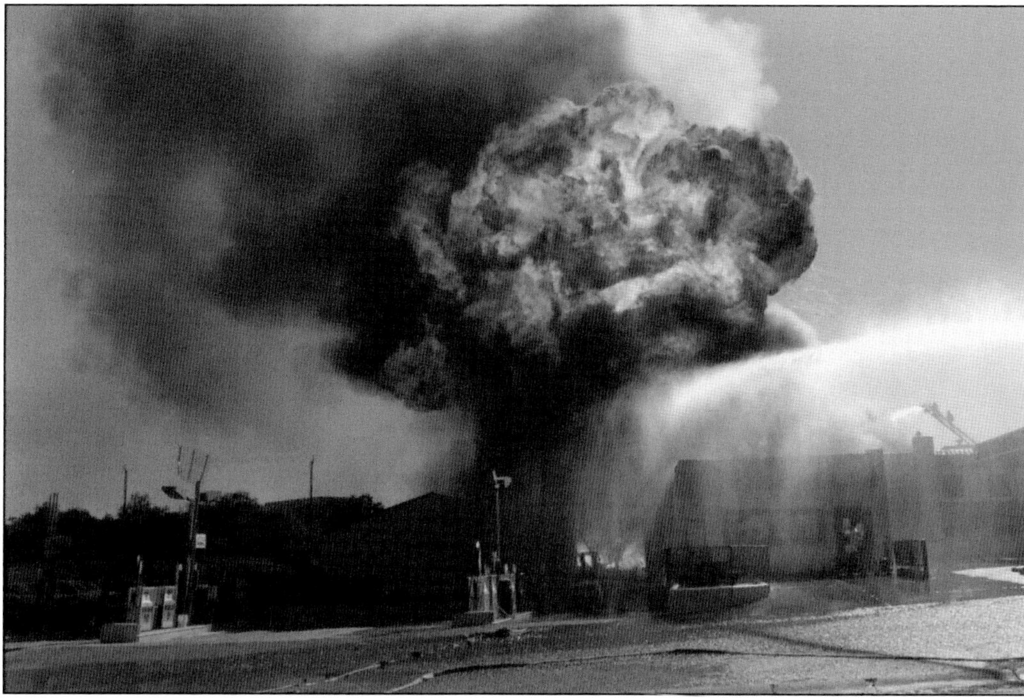

The Bennett Fire on September 1, 2005, occurred when high octane fuel was being transferred between two containers without proper grounding. A static spark set off the fuel and fire quickly spread throughout the building and its drums of fuel. The aircraft rescue engine was dispatched to the fire with its 200-gallon foam supply. Oily runoff had to be stopped before it made its way to nearby Granite Creek. (Both, courtesy of Les Stukenberg.)

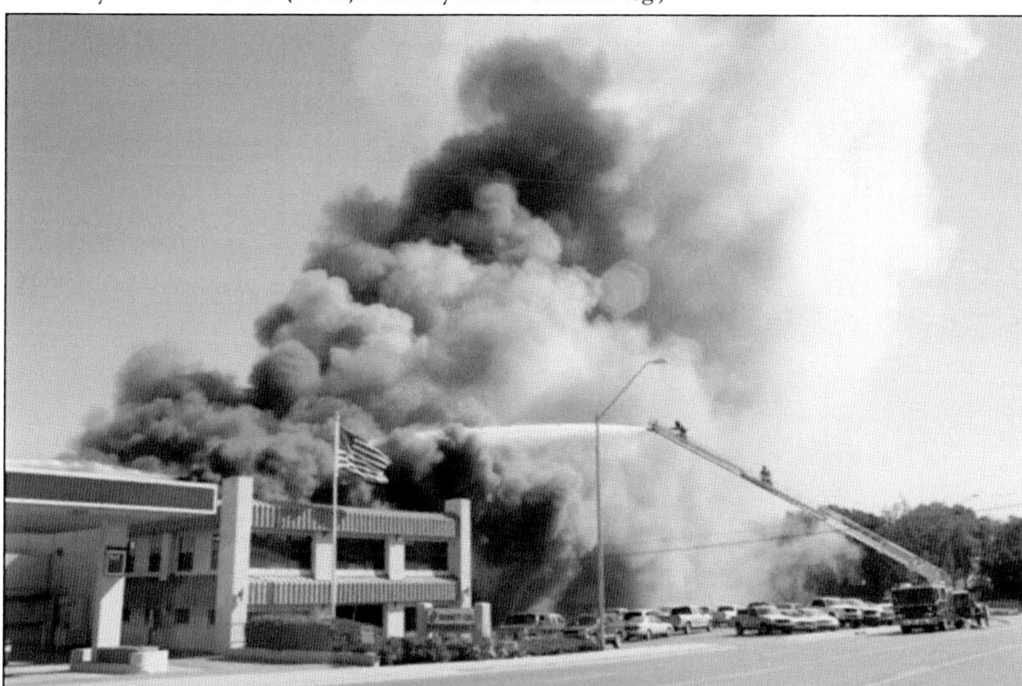

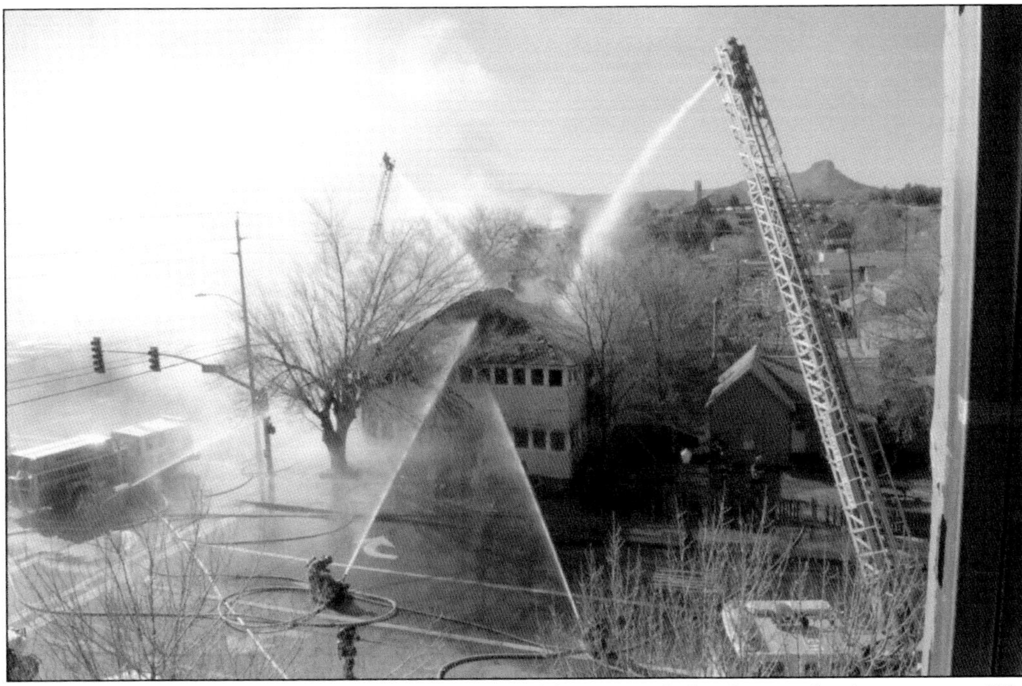

Started by an unattended pan of frying oil, this blaze gutted the Coronado Apartments. The building was originally built between 1910 and 1924 by the Ehle family. The Ehles had been in Prescott since 1864 and managed multiple hotels. They also helped build the log governor's mansion now located at Sharlot Hall Museum. Fighting the fire proved difficult due to all the adjacent power lines blocking the aerial hose streams. Power was cut to a large portion of the downtown area initially in order to keep the scene safe for firefighters. (Both, courtesy of Les Stukenberg.)

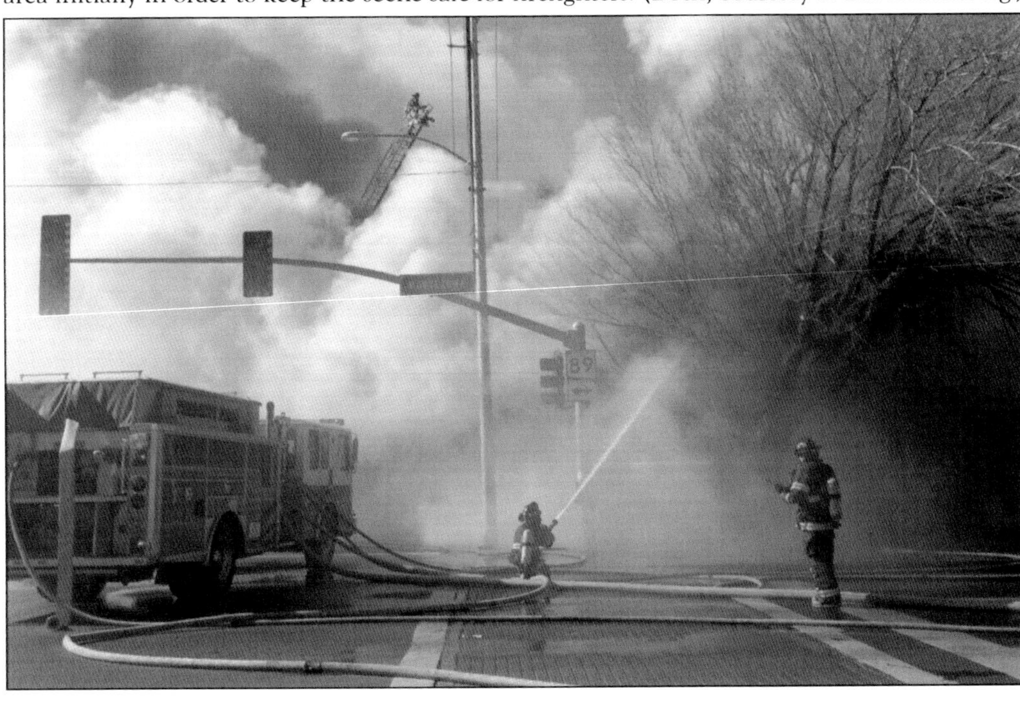

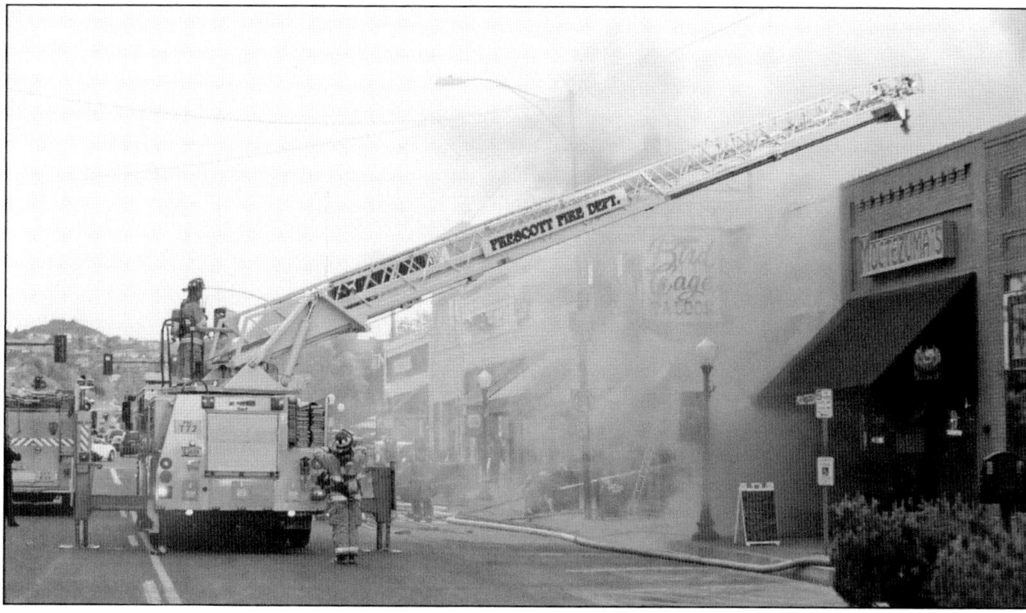

Starting on May 8, 2012, in the middle of historic Whiskey Row, some 50 firefighters from three agencies, nine engines, and three aerials would be required to stop a fire that conjured images of the 1900 fire. The fire destroyed three businesses, including the famous Bird Cage Saloon and its stuffed bird collection. It took nearly three hours and some 750,000 gallons of water to get the fire under control. Damage was estimated at $1 million to the buildings and another million to the contents. (Both, courtesy of Matt Hinshaw.)

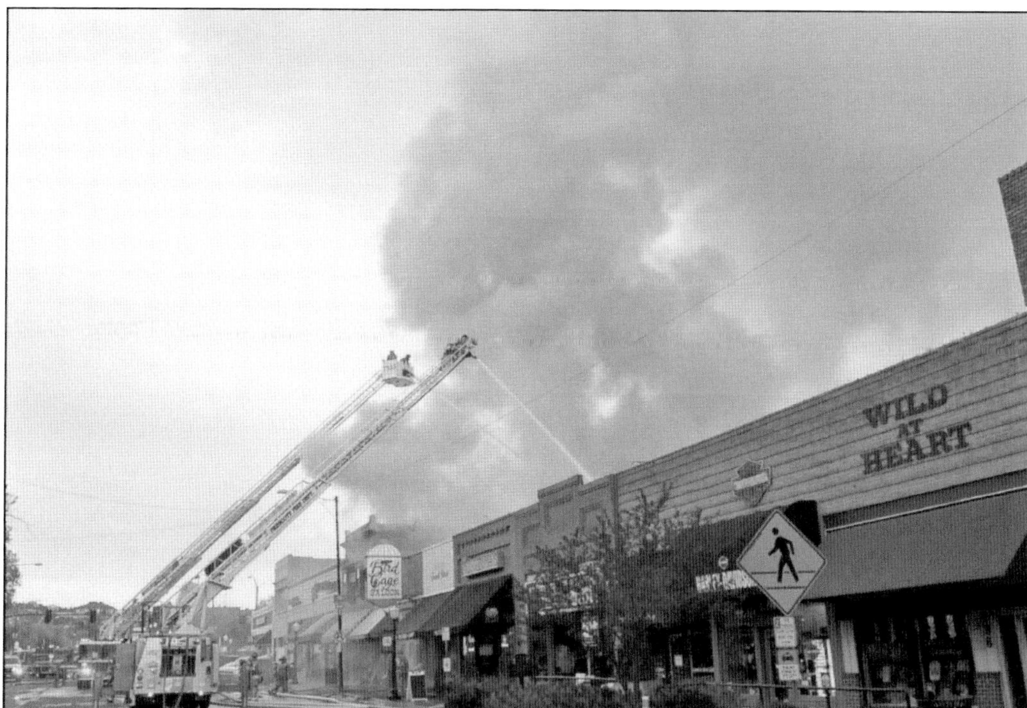

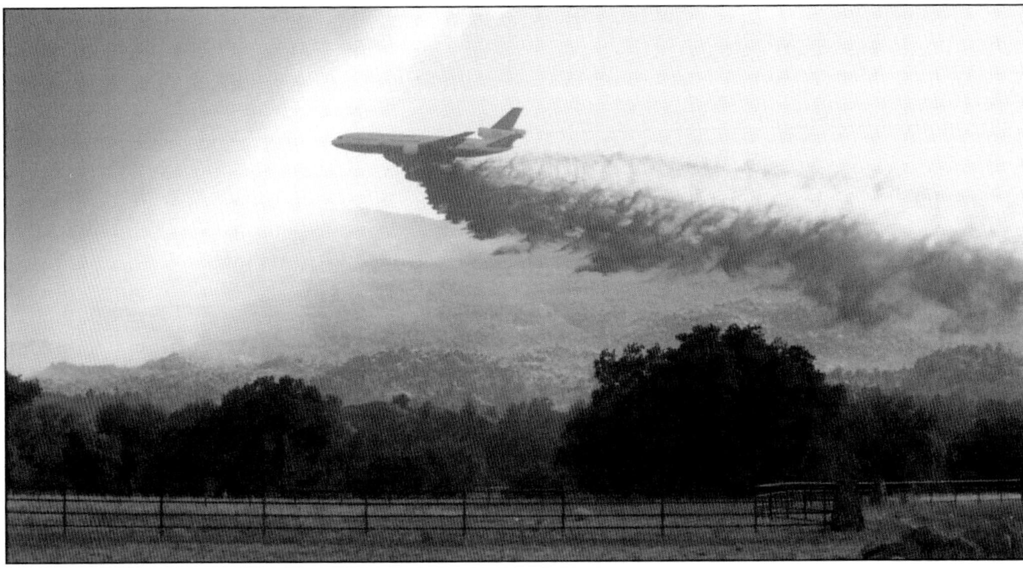

The human-caused Doce Fire started on June 18, 2013. It would consume almost 7,000 acres by its end, scorch the better part of Granite Mountain, and cost some $7 million dollars to fight. The coordinated training efforts of the local fire agencies again garnered results. The heavily impinged communities of American Ranch, Sundown Acres, and the Williamson Valley as a whole were protected and not one structure was lost. The Granite Mountain Hotshots gained recognition on the fire when they saved a 1,000-year-old, 14-foot-diameter alligator juniper from burning on their namesake mountain. (Both, courtesy of Les Stukenberg.)

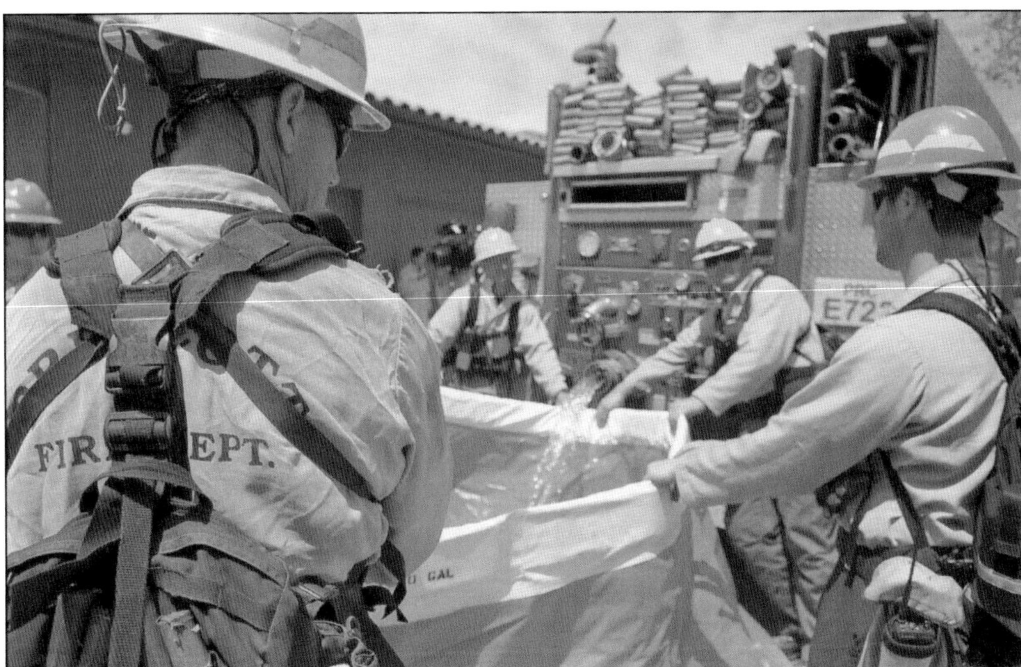

Six

Special Operations and Practices

Like many other fire departments in the nation, the scope and service of the Prescott Fire Department has grown over the years. Fire brigades came to be seen as a standing support service for their community. Fire prevention efforts became a standard in an effort to stop fires before they began. Medical training and certification was added so firefighters could render medical assistance immediately. Eventually, changing technologies warranted further specialization of firefighters' skills in aircraft firefighting, technical rescue, and hazardous materials.

Currently, Prescott Fire Department has an active fire prevention division, which sees to the proper coding of all new buildings as well as assigning the fire crews in their annual fire inspections of existing buildings. Additionally, firefighters conduct educational prevention training with elementary schoolchildren and firefighter training with high school students through the Fire Explorer program.

Earnest Love Airfield is one of the busiest airfields in the state due to the presence of Embry Riddle Aeronautical University, as well as frequent commercial and private flights. As a result, the department is required to maintain an Aircraft Rescue and Firefighting certified engineer at the airport during normal operating hours. Additionally, the Federal Aviation Administration has provided a specialized engine, formerly called a crash truck, to be stationed at the airport. This vehicle is manned even when the remainder of the crew at the station responds to other nearby emergencies.

Two of the more highly specialized operational teams on the department are the technical rescue and hazardous materials teams. Select members of Prescott Fire Department's firefighters serve on these regionally functioning teams, working together with firefighters from Central Yavapai, Chino Valley, Groom Creek, and Mayer Fire Departments, on low frequency, high risk events. The technical rescue team regularly removes injured hikers and bicyclists from Prescott's multitude of trails or conducts pickoffs from cliff faces and flooded waterways. The hazardous materials team has been used to stop leaking vessels, identify unknown chemicals, and evaluate gas concentrations in endangered buildings.

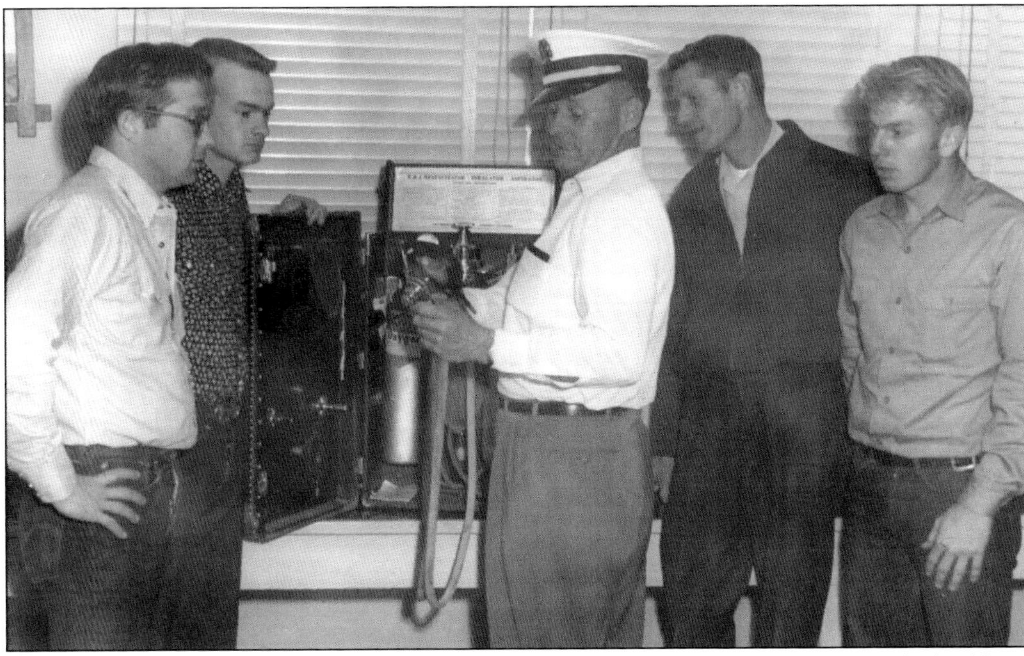

The Inhalator, pictured above in the hands of Chief Jack Custer and in use below, was the backbone of EMS calls all the way back to 1930. When the department suggested purchasing the unit in 1929, the city stated there was no money in the budget for the device. The firefighters voted to sell one of their $1,000 Liberty Bonds in order to make the purchase, on the condition that the city would allocate money in the coming year and pay back the firefighters. The Inhalator was once credited with saving the life of a two-week-old baby by running for 116 hours straight.

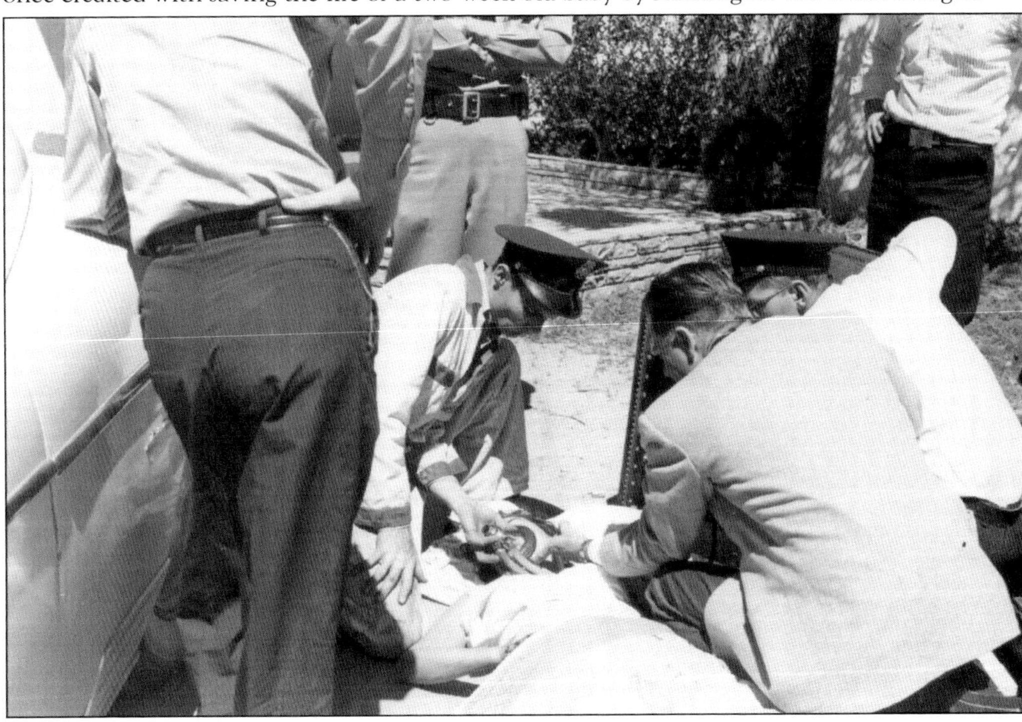

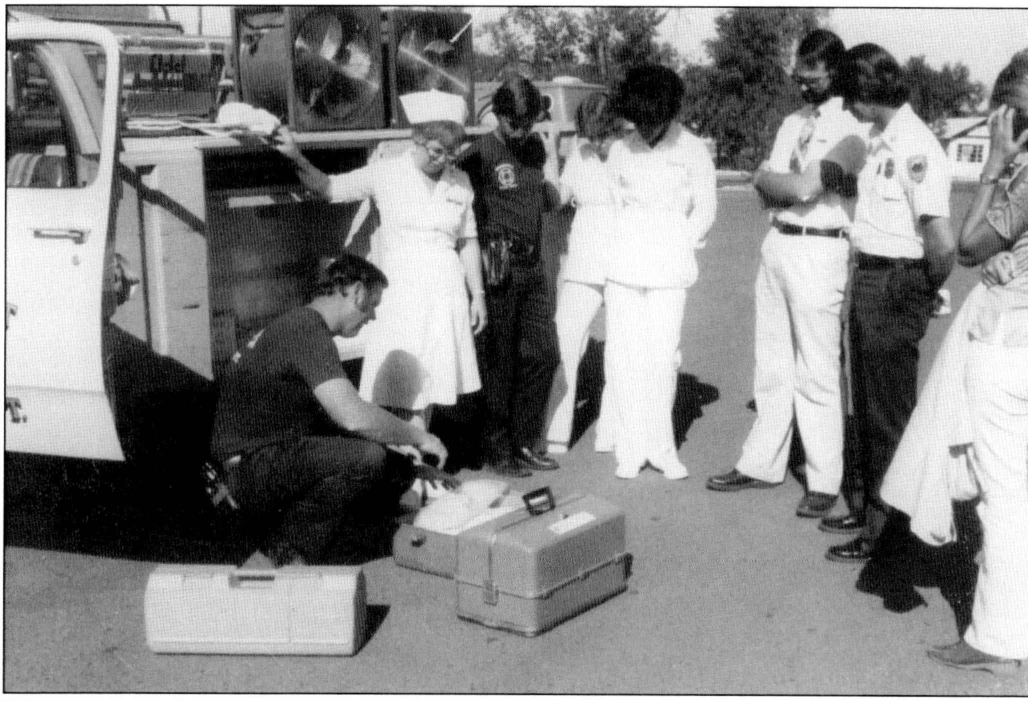

While initially firefighters served as first responders by circumstance, with only basic first aid knowledge, they soon advanced to Basic Emergency Medical Technicians (EMT-B). In 1979, six Prescott firefighters took an even more advanced course, becoming Intermediate Emergency Medical Technicians (EMT-I). Yavapai Regional Medical Center, then a county hospital, had to keep a doctor available at all times for the medics to contact for clearance to administer medications and their staff had to be educated on what skills EMT-I provided in the field. Today, all Prescott engines now operate with at least one Certified Emergency Paramedic (CEP) on the engine. All new hires have to have their EMT-B certification prior to hire. The last firefighter hired without an EMT-B certificate came on board in 2001. He was given one year to complete the program or be fired.

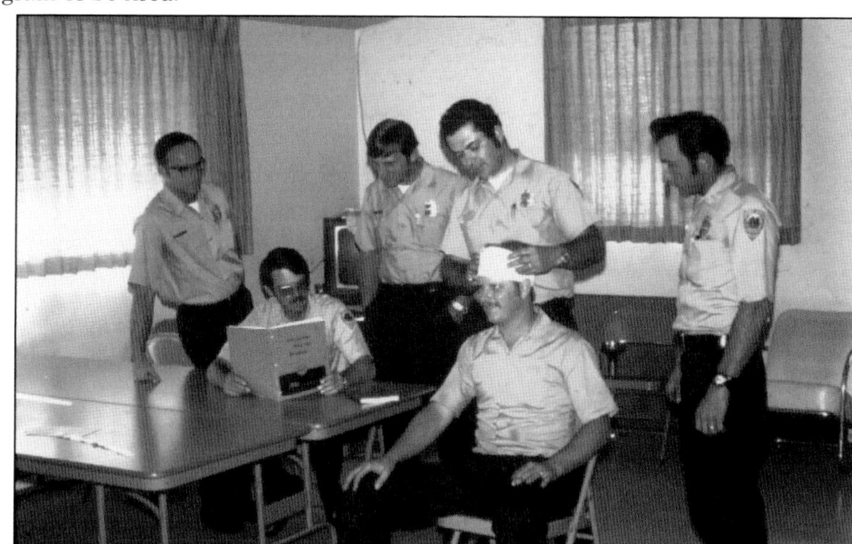

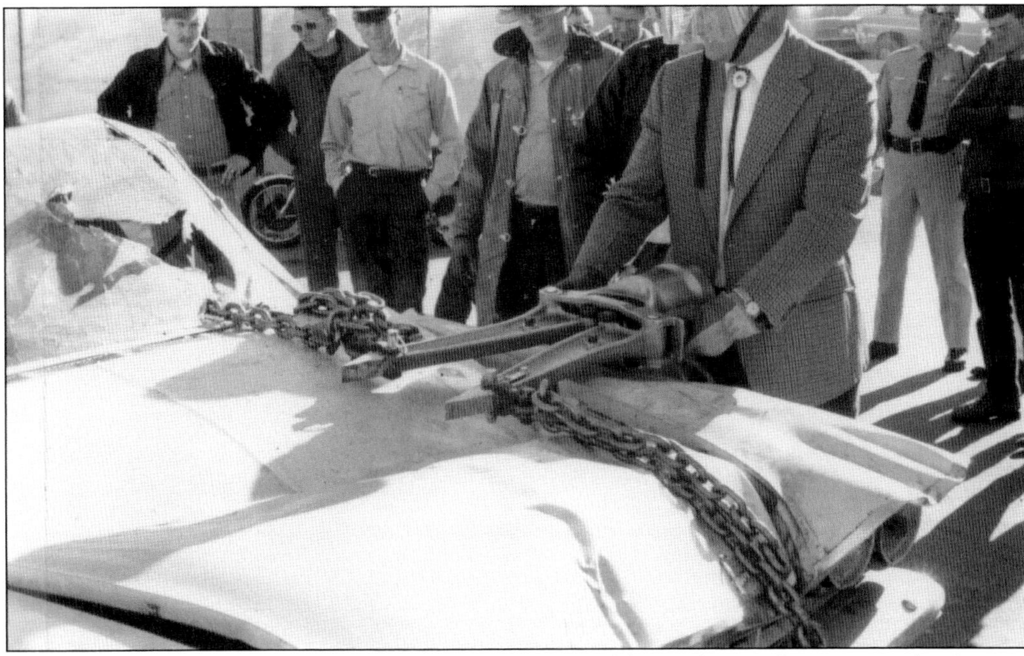

A salesman demonstrates the use of a Hurst tool for possible purchase. The woman who donated the vehicle in the photograph returned several days after this demonstration with a change of heart and requested the return of her vehicle. The department sadly informed her that its return was no longer an option.

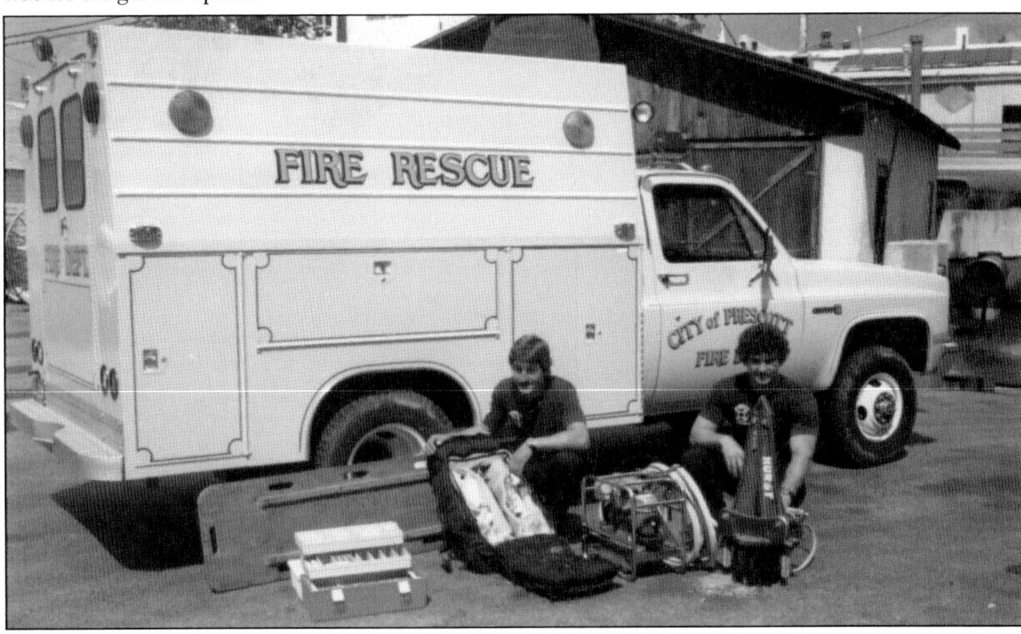

Dave Scalia and Virgil Esmont are showing off some of the materials and equipment stored on Rescue 71. The Rescue 71 vehicle was purchased prior to the EMT-I medics finishing their medical training in anticipation of the new service that would be provided to citizens. The Hurst tool that Esmont is holding was purchased by the Department of Public Safety on the condition that the rescue team respond to vehicular emergencies anywhere in the region.

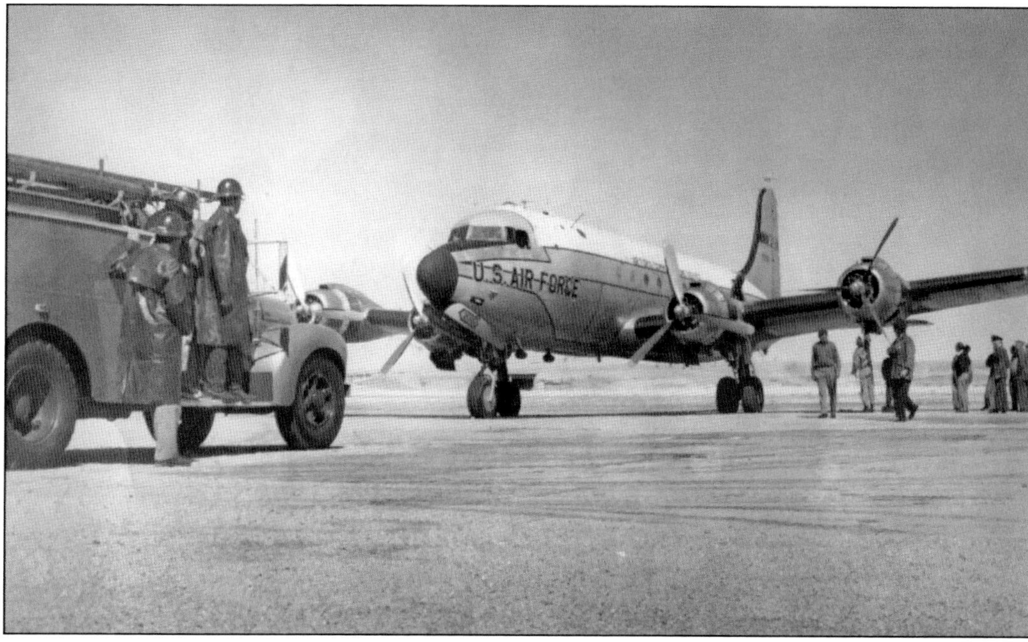

Above, an Air Force C-154 made an emergency landing at Earnest Love Field in 1962 after reporting smoke coming from the left outboard engine. The fire had extinguished itself by the time it landed and all 12 crewmembers aboard were unharmed. Below, Foam 73 sits nose-to-nose with the 1945 B-17G Flying Fortress known as *Sentimental Journey* in 2009. The bomber landed in Prescott as a part of a promotional and fundraising tour. The Aircraft Rescue Firefighting (ARFF) component of Prescott Fire Department keeps a dedicated ARFF certified engineer on Foam 73 during the normal operating hours of the airport. The presence of Embry Riddle Aeronautical University in Prescott makes Earnest Love Field one of the busiest airports in the state due to the students' frequent touch-and-go traffic.

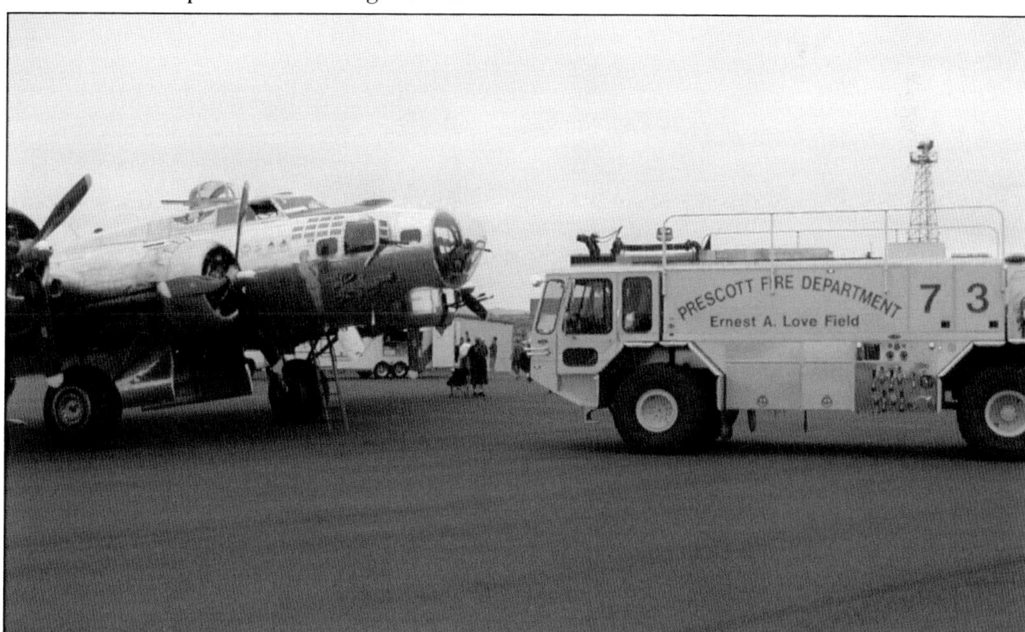

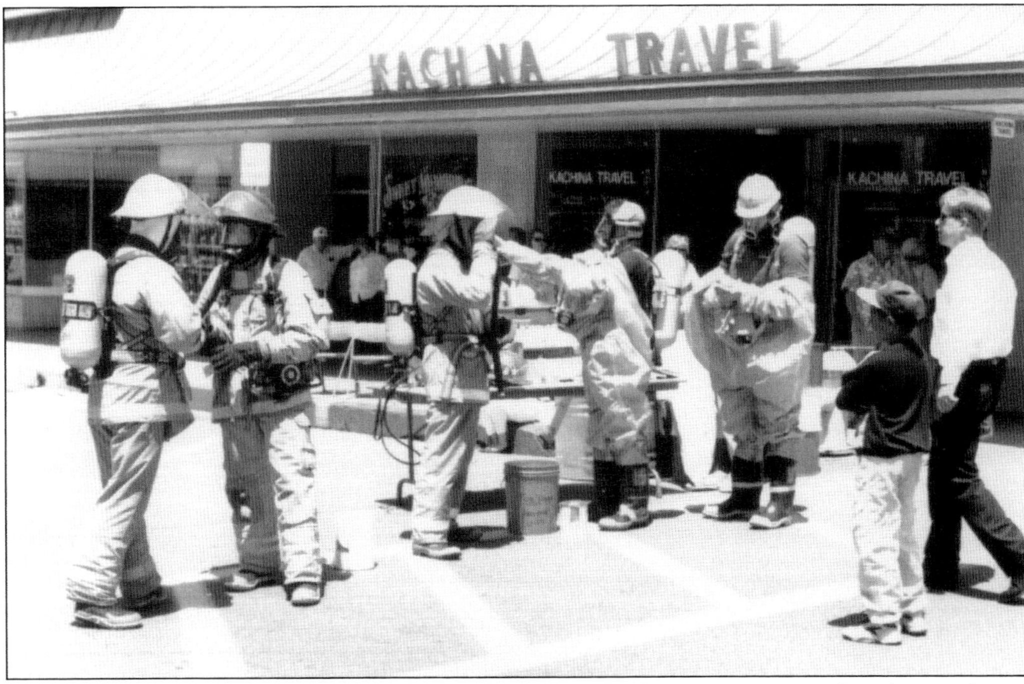

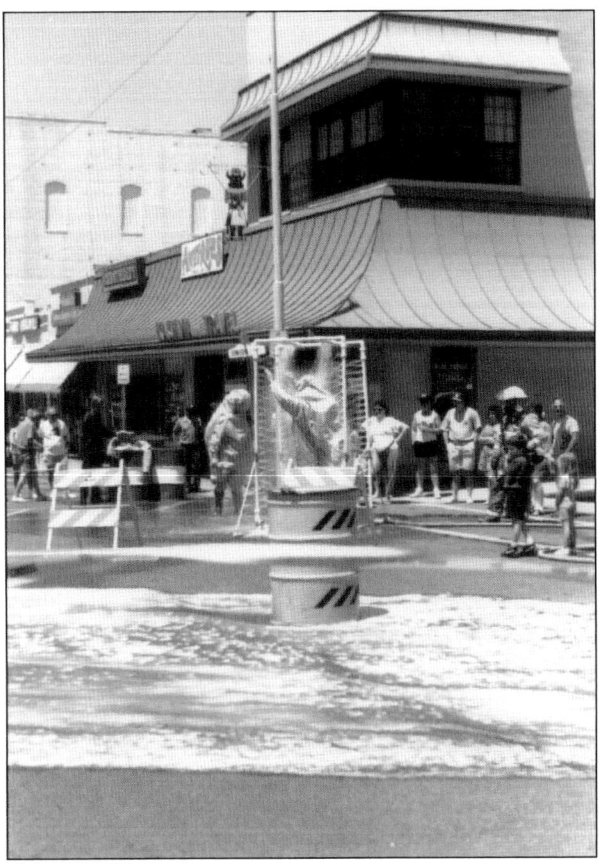

To improve citizens' understanding of the various technical skills required by the department, annual public displays and demonstrations have been utilized. Seen here, members of the hazardous materials team conduct a simulation of stopping a leak and taking samples of materials to be tested for identification. The hazardous materials team is a joint regional team run with hazardous materials technician firefighters from both Prescott Fire Department and Central Yavapai Fire District. In 1974, Prescott firefighter Denny O'Reilly was tasked by Chief Hamilton with educating the department on hazardous materials as O'Reilly was a trained teacher with a degree. Today, hazardous materials technicians obtain their state certifications through a month-long program and train together monthly to maintain their skills.

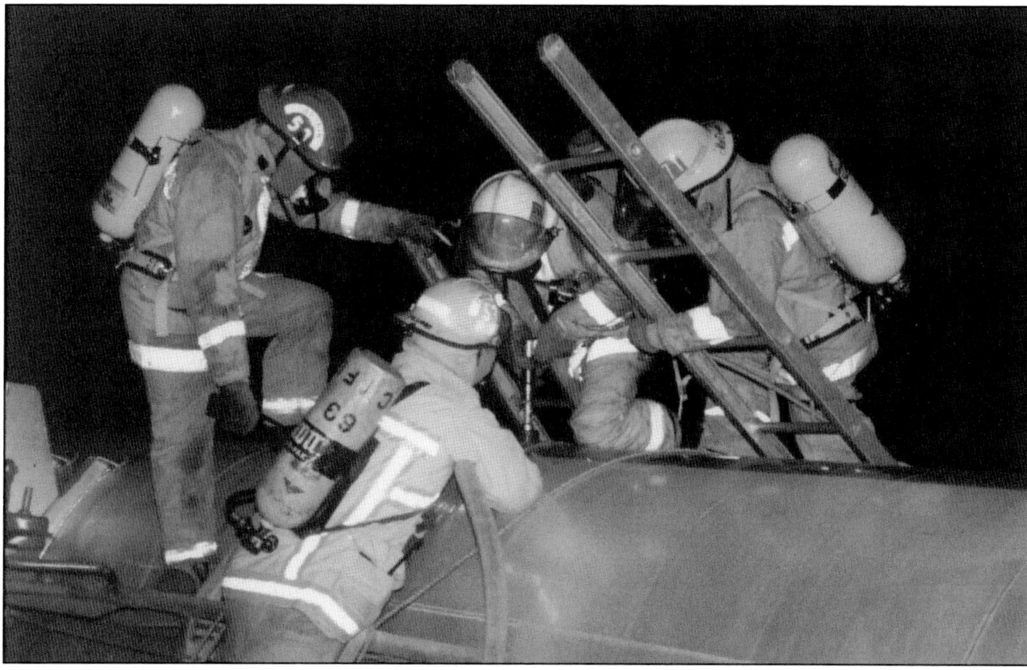

The hazardous materials team has been called upon regularly to offload liquids from transport tankers that have rolled over. Often, the tankers are carrying loads of fuel for distribution. Excessive speed and roundabouts have both been contributors to the rollovers. Equipment and materials for these kinds of responses are housed at Station 75 and on a special hazardous materials vehicle there, which started its life as a Mac Tool truck before being reconfigured for use by the team. Station 75 has been alternately manned by Prescott and Central Yavapai Fire Districts since its construction in 1993 and is always staffed with at least one hazardous materials technician.

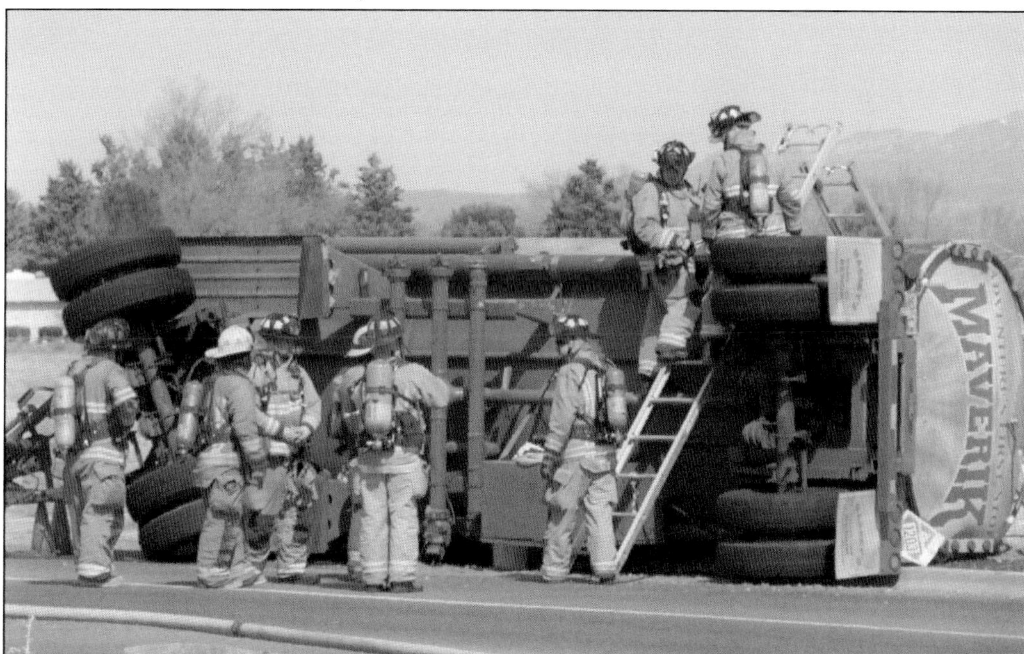

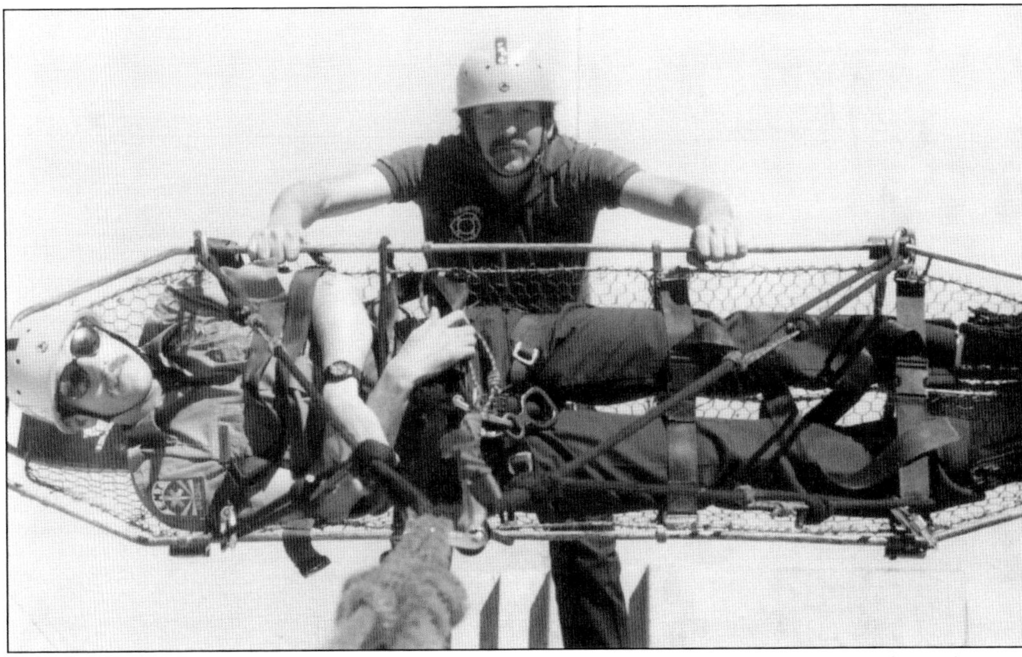

Above, Brad Malm lowers a "packaged" Jay Fillingim in a Stokes basket. The technical rescue team is a regional team with members hailing from not only Prescott Fire Department, but also from the fire agencies of Central Yavapai, Chino Valley, Mayer, Groom Creek, and Williamson Valley, as well as members of the Yavapai County Sheriff's Office backcountry team. The Prescott component of the team began in earnest in the late 1970s and slowly developed into the program it is today. Members of the team drill on a monthly basis and conduct large-scale multi-faceted drills on a quarterly basis. Below, the need for such training can be seen as Dave Scalia oversees the descent of a rock climber with a broken leg.

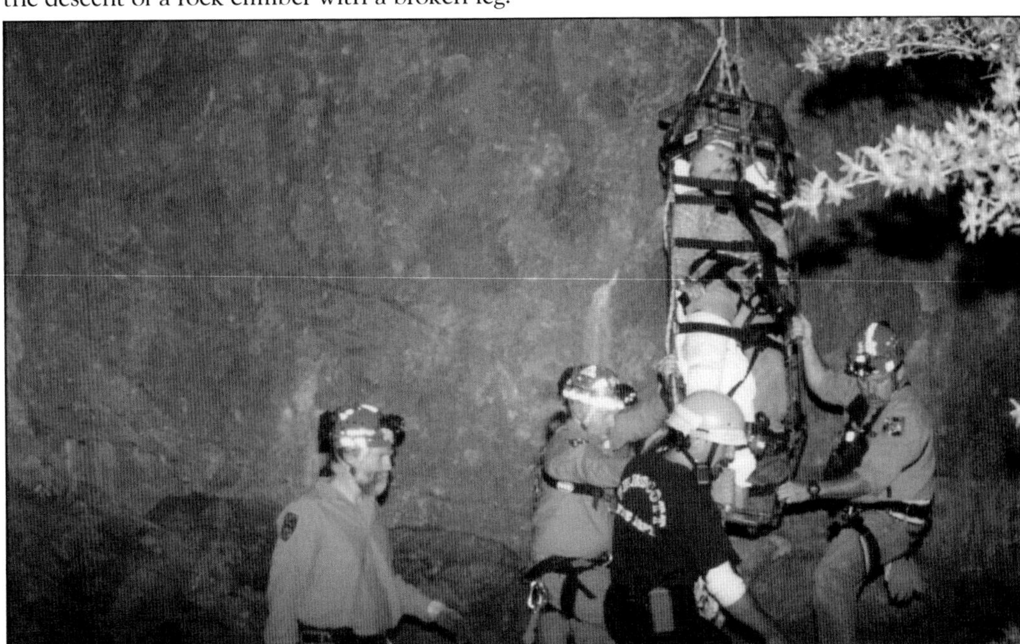

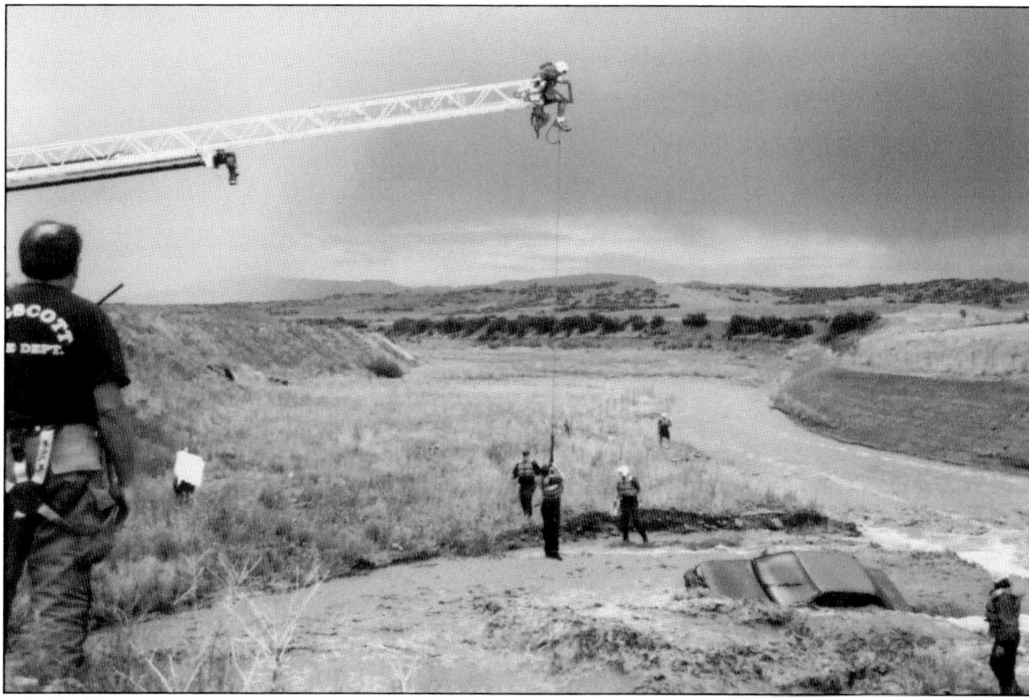

The ladder company is pictured rescuing a driver from the roof of their rapidly sinking car. Local monsoon flooding of otherwise dry riverbeds leads to annual rescues of folks unfamiliar with the hazards associated with Prescott's low water crossings. Both of the department's current ladder companies have a special "cinch" ring device used for rapid deployment on this type of rescue.

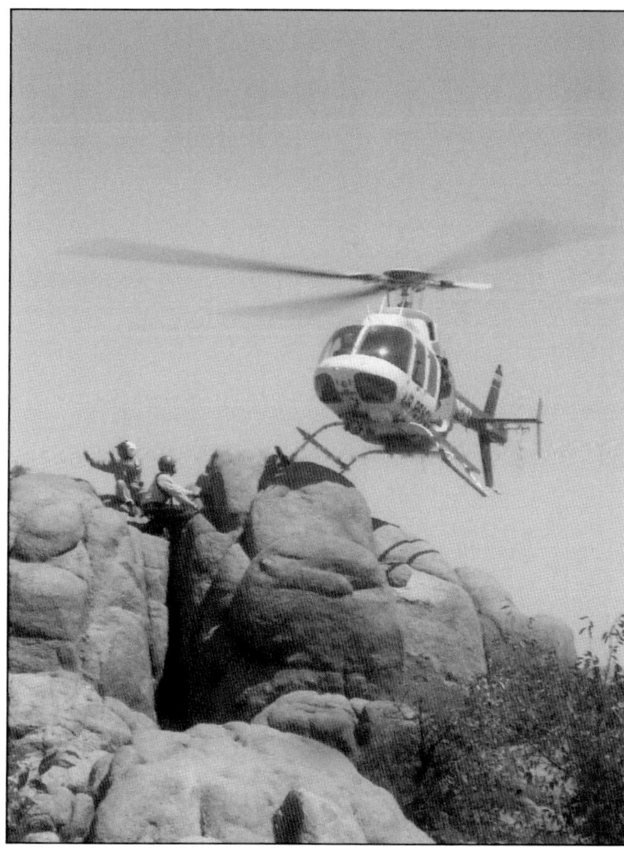

Due to the difficult access to many of Prescott's scenic areas, local technical rescues often make use of helicopters. Here, regional team members conduct their annual helicopter insertion and extraction training at Watson Lake. The team's members also conduct training relating to confined space rescues, swift water rescues, and high angle rope rescues.

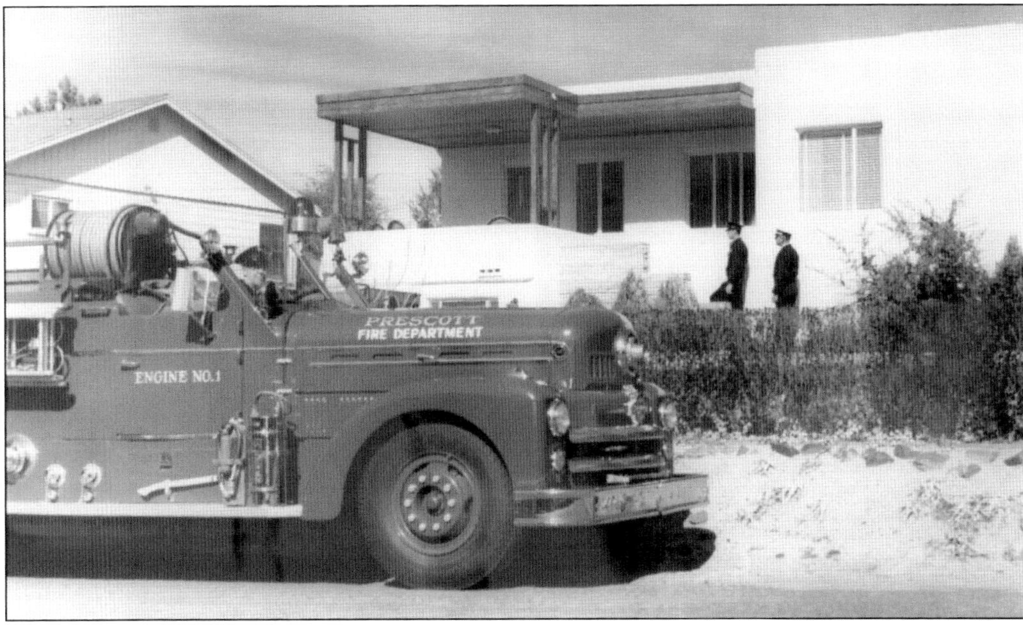

In January 1958, Prescott Fire Department started offering voluntary home inspections. Firefighters would walk through neighborhoods, going door-to-door and asking homeowners if firefighters could inspect their home. Besides decreased house fires, two ancillary results came from the program: firefighters were issued formal uniforms so citizens would know they were in fact with the fire department and a two-way radio was installed in the engine so it could respond to fires while out of quarters. One firefighter remained with the vehicle to monitor radio traffic. The firefighters' presence in homes also fostered a new generation of youth interested in firefighting, and the department developed a training program to expose them to firefighting tactics and first aid.

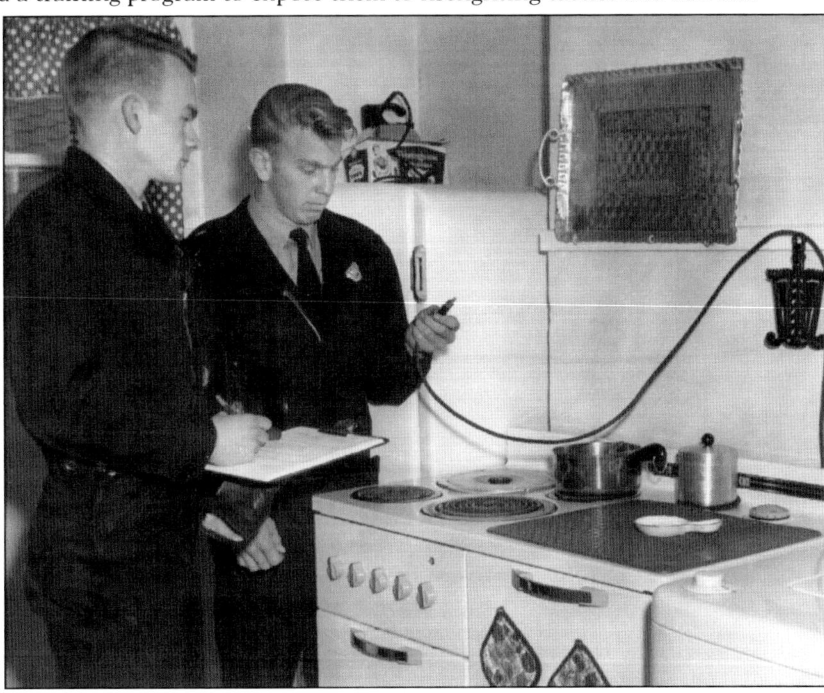

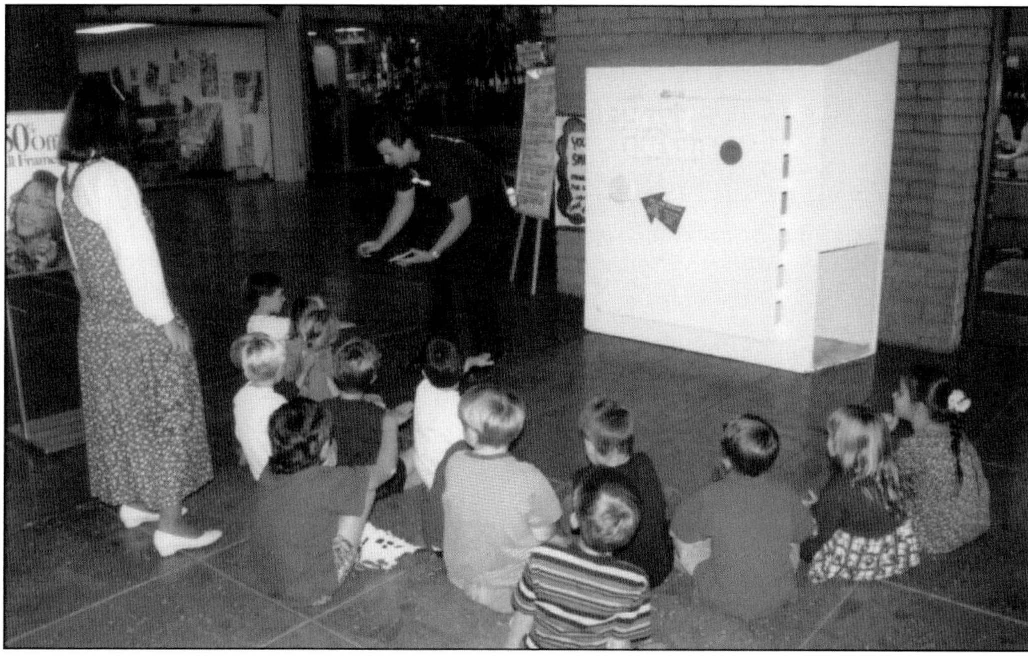

In the 1980s and 1990s, the public education and prevention program for children moved into Ponderosa Mall. Elaborate props were constructed to help create visuals for the elementary students who were brought to the mall for the program. Students were able to walk through an oversized smoke detector, oversized hot faucets, and even a bedroom mockup, which demonstrated ways to escape a fire out the window of a bedroom. Currently, engine companies visit with first-grade classes in their classrooms to discuss topics like staying low in smoke and other Learn-Not-To-Burn topics.

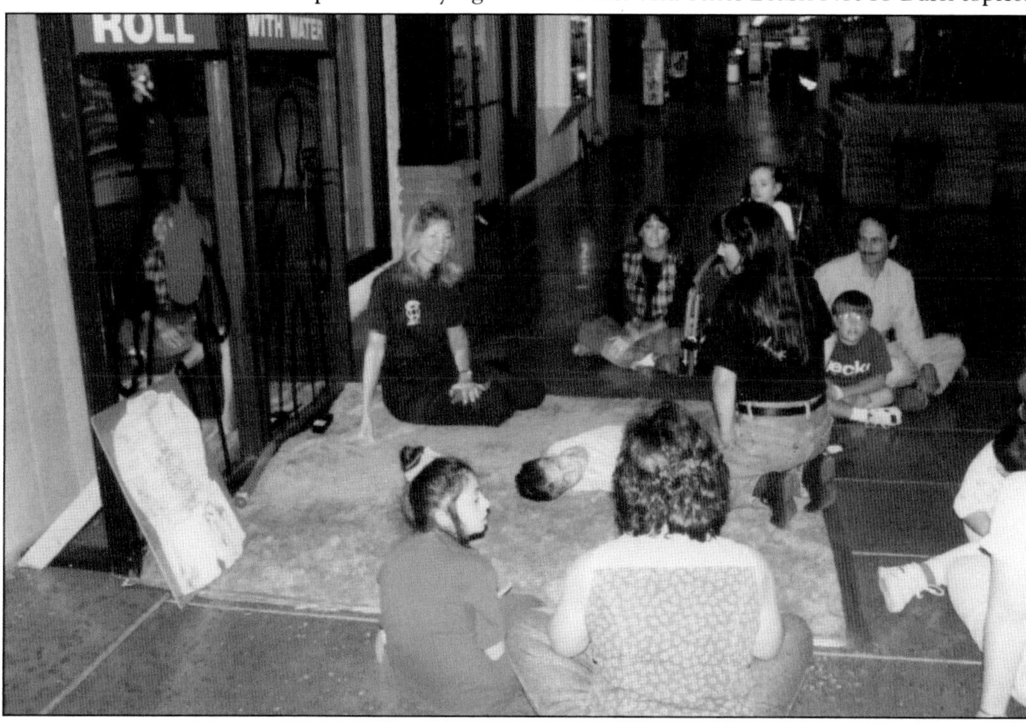

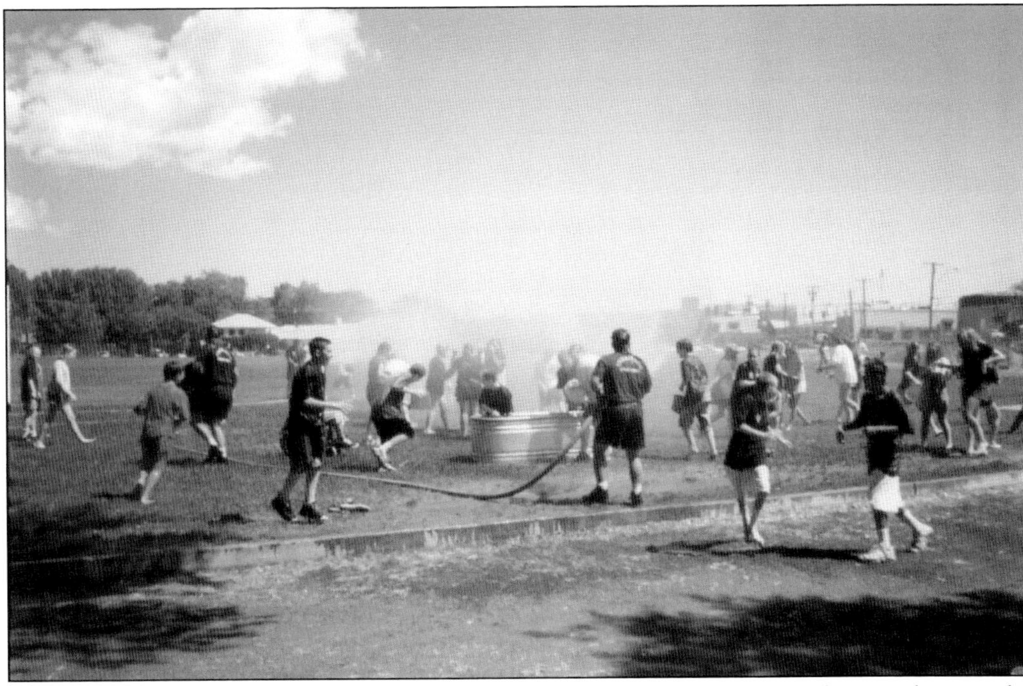

Prescott Fire continues its contact with middle school–aged students through annual career day presentations as well as this activity, called Fire Trap, where students engage in firefighter-related activities, with the event ending in a massive water fight.

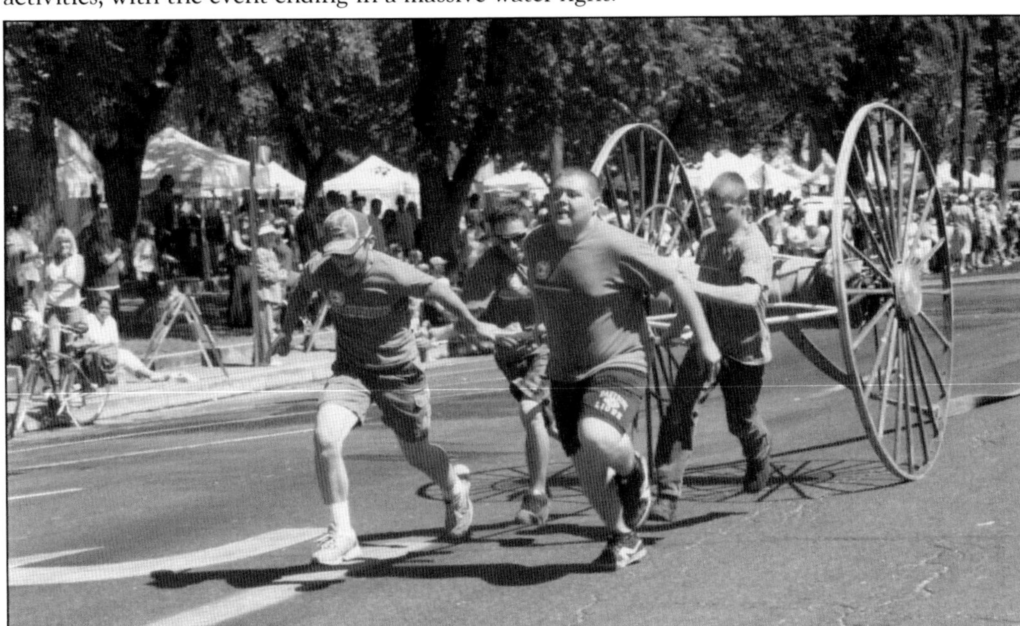

At the high school level, students interested in firefighting as a possible career can join the department's Fire Explorer program. The program was created in 1997 and exposes students to the various skills associated with the profession. From left to right are Fire Explorers Ty Seets, Daly Parra, Michael Williams, and Peter Hriljac, competing in the annual hose-cart races.

Seven

WILDLAND

The city of Prescott is surrounded by one of the largest Ponderosa Pine forests in the nation. Prescott National Forest has managed this resource. The former forest management system was suppression-minded, with all fires being snuffed out as quickly as possible. This allowed the forest around the community of Prescott to accumulate thicker and denser fuel loads. The frequency and intensity of fires increased as the years progressed. Compounding the risk was Prescott's annexation of surrounding county lands, which had also accrued heavy vegetative growth. Residents were building more frequently in areas primed for a destructive wildland fire.

By the 1960s, it was recognized that the firefighters should have training specifically directed towards this increasing wildland threat. Prescott firefighters increased their knowledge through training as well as by developing a wildland team that would send engines to fight wildland fires in distant communities and states. Knowledge and skills learned abroad were then brought back to the Prescott community.

In 1999, Prescott was designated one of the highest risk wildland-urban interface communities in the nation. Then-chief Darrell Willis and the battalion chiefs began generating a vision in the following years to not only develop the department's wildland response team, but also create a preventive wildland resource in having a fuels reduction crew as a part of the department.

This vision inched forward one step at a time, beginning in 2001 with a grant-supported fuels mitigation crew who would remove heavy fuel loads from around homes in Prescott. As the skill and training of the crew grew, their ability to take on more challenging assignments increased. By 2004, they were recognized as a Type 2 Initial Attack crew and named Crew 7. They achieved their goal as a nationally recognized Interagency Hotshot Crew in 2008, and formally adopted the name of the Granite Mountain Hotshots, identifying with the mountain that serves as a backdrop for Prescott. The crew saw its tragic end on the Yarnell Hill Fire on June 30, 2013, when shifting winds caused the fire to overrun their position, killing 19 of them.

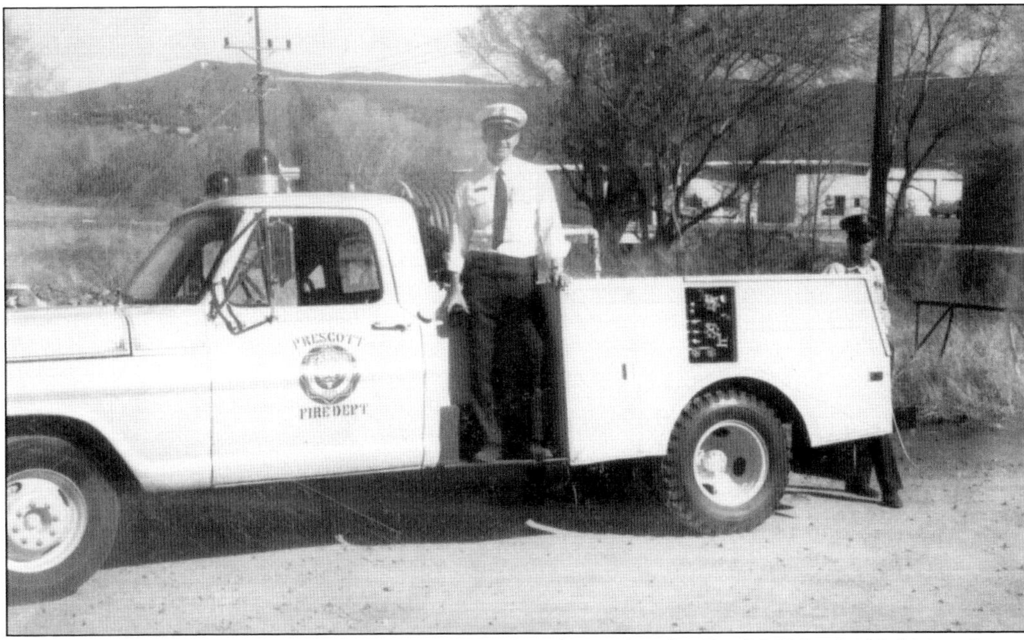

Bob Hinshaw stands on the newly constructed brush truck. The Ford 250 had 130,000 miles on the engine when it was retrieved from a scrap yard and was amended with scrap materials to create a tank and storage cabinets. Assistant Chief Jim Murphy stands at the tailboard.

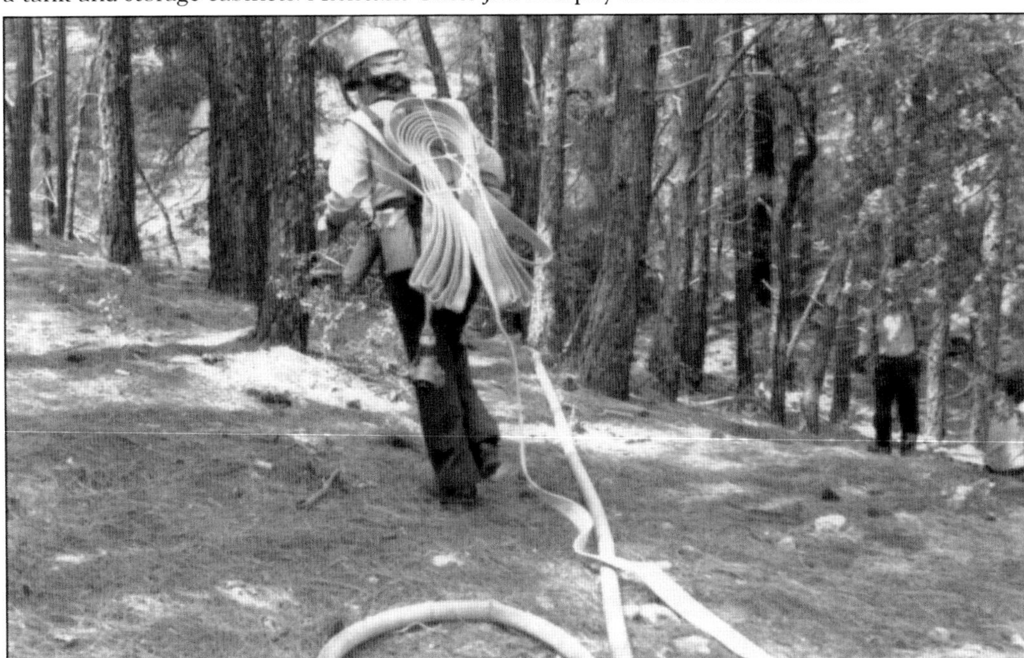

The traditional Gasner pack shown here is comprised of 100 feet of one-and-a-half-inch hose followed by a gated wye and 100 feet of one-inch hose with a nozzle; it is used for rapid deployment of a progressive hose lay. The new generation Travis pack used locally has the same complement of hose and appliances but at half the weight, making it less cumbersome. The redesign was created by Prescott National Forest firefighter Bob Travis.

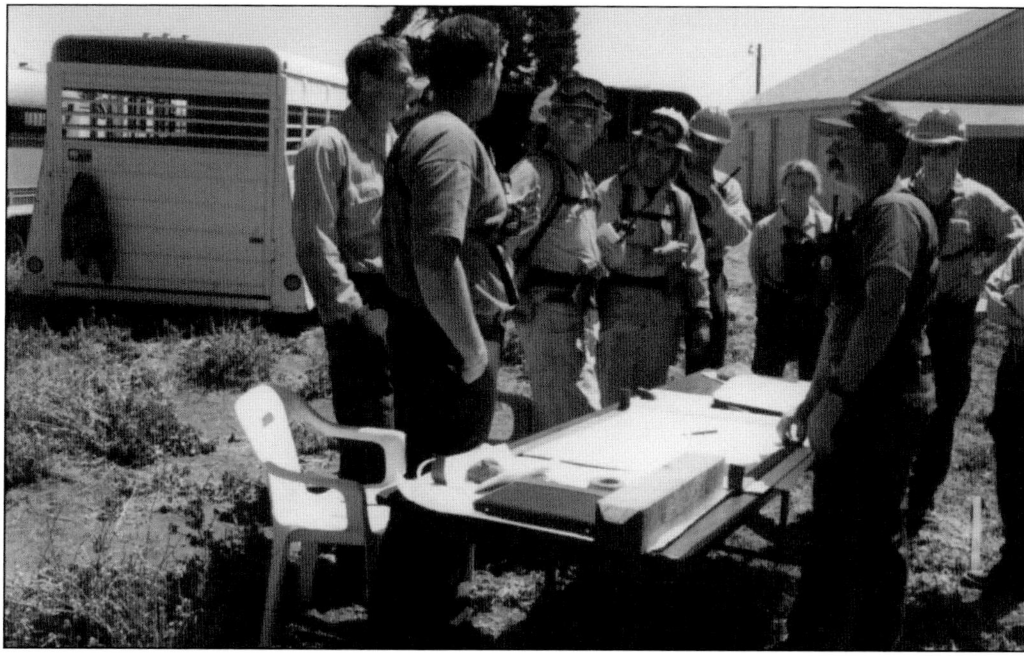

Local fire agencies in the Prescott area have annually conducted joint wildland training drills since the 1990s called the Basin Drill. This ongoing training has resulted in smooth mutual aid responses and fosters familiarity with each agency's resources and capacities. Pictured with goggles on their helmets are Prescott firefighters Tim Sheehan and Allen Snyder.

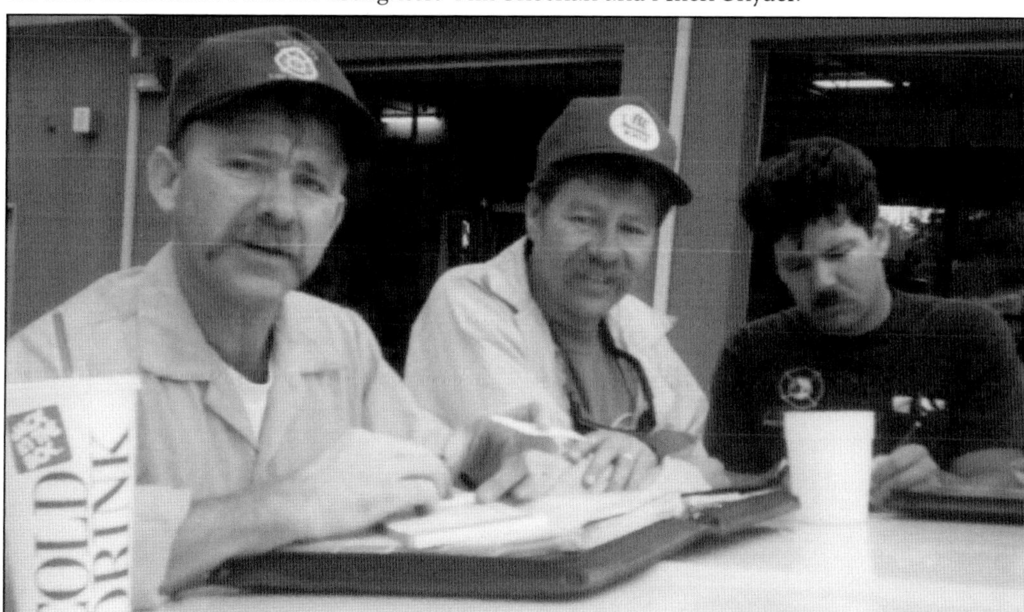

From left to right, firefighters Scott "Kermit" MacDermid, Marty Cole, and Don Banning are pictured preparing materials for a Basin Drill while serving on a Type Six patrol. Marty eventually worked his way through the wildland ranks to become a safety officer on a national incident management team, also serving as a captain on the Granite Mountain Hotshots in 2004 and as their superintendent in 2005.

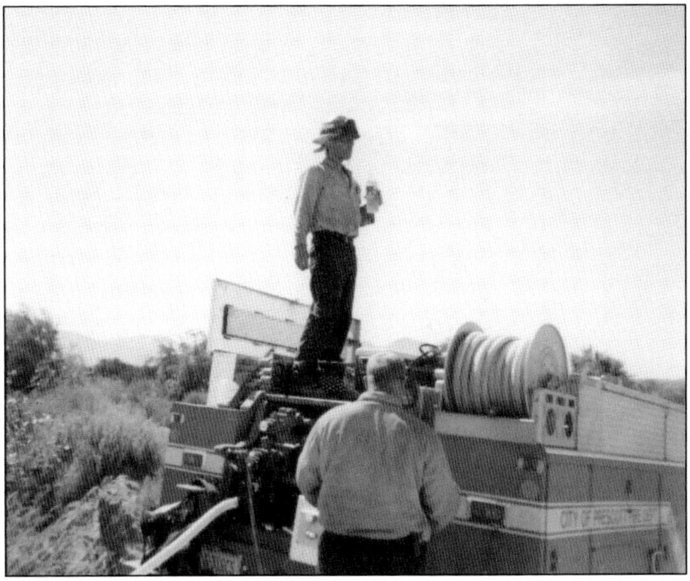

Ralph Lucas stands on top of a Type Six patrol for a better view of the Aguila Fire with Troy Steinbrink at the pump panel. Prescott Fire has three Type Six patrols and two Type Three engines available for dispatch around the country. Additionally, Prescott fire has sent Type One engines to assist with structure protection on fires in southern California. In addition to the engines, multiple members of Prescott Fire Department have served in positions on national incident management teams.

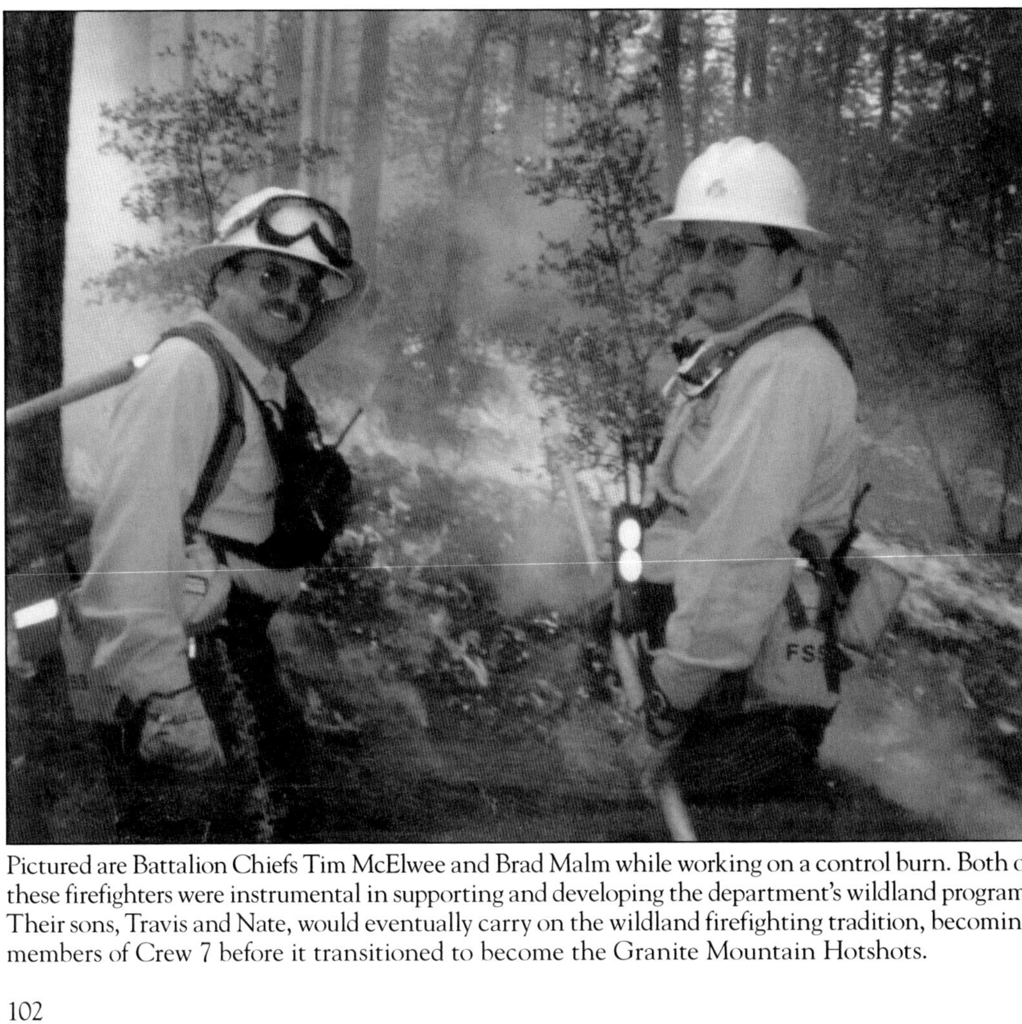

Pictured are Battalion Chiefs Tim McElwee and Brad Malm while working on a control burn. Both of these firefighters were instrumental in supporting and developing the department's wildland program. Their sons, Travis and Nate, would eventually carry on the wildland firefighting tradition, becoming members of Crew 7 before it transitioned to become the Granite Mountain Hotshots.

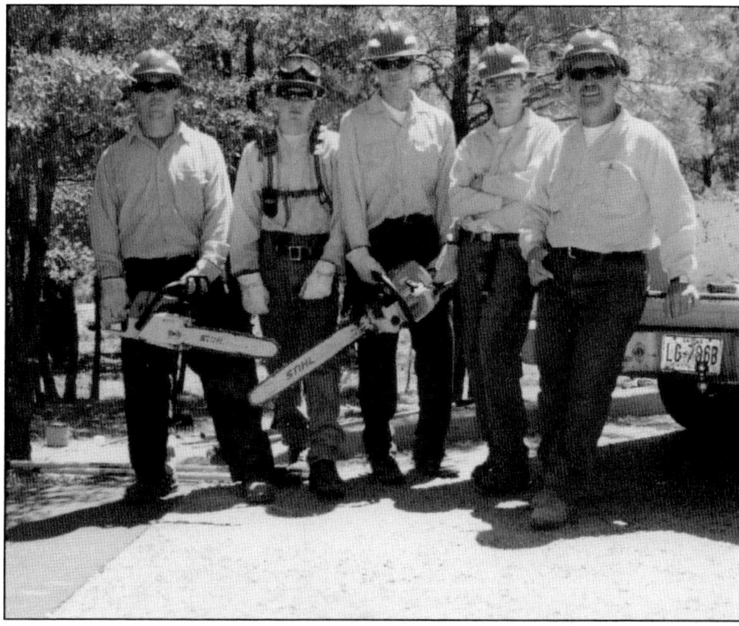

The Granite Mountain Hotshots' humble beginning was with the inception of a grant-funded fuels reduction crew. The fuels crew, Crew 7, served a proactive function in the city, reducing fuel loads from around homes. In its first year, it created defensible space around 392 homes. Pictured here from left to right are Crew 7 members Brian Padilla-Melton, Travis McElwee, Chris Porter, Cody Hollenshead, and Ralph Johnson.

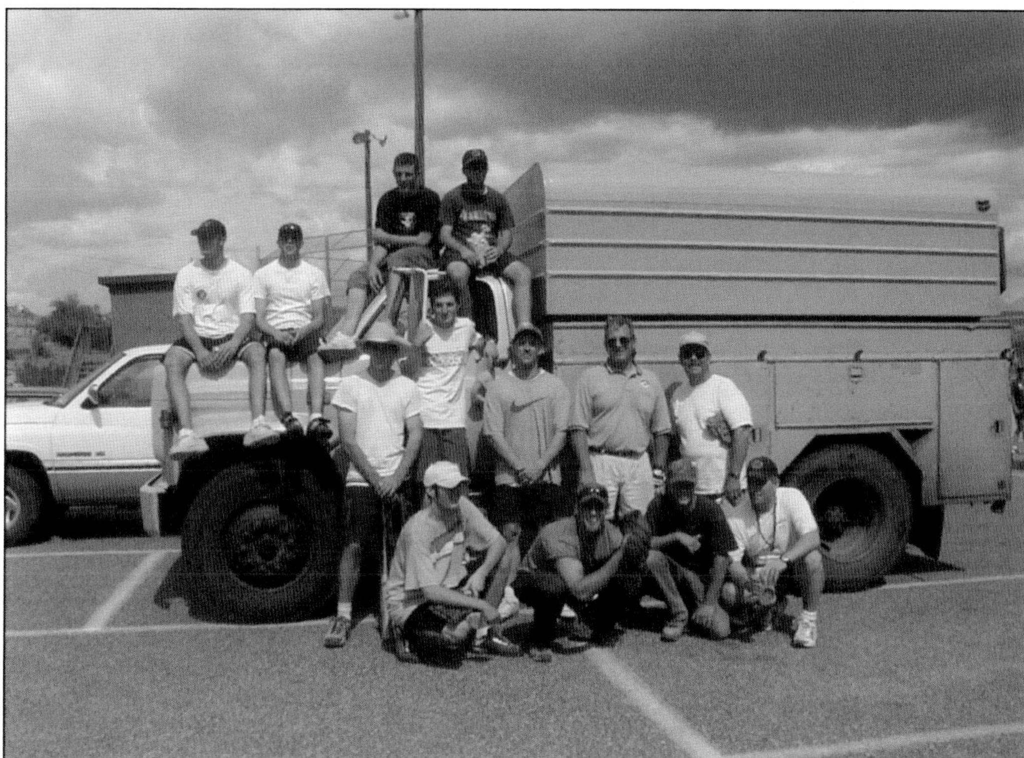

A year-end softball match became a tradition for the crew from their first year, probably due to their softball-loving leader, Duane Steinbrink. From left to right are (first row) Tom Donovan, Mike Huddleston, Josh Quinn, and Bill Hickey; (second row) Ben Huddleston, Jared Willis, Brian Padilla-Melton, Duane Steinbrink, and Ralph Johnson. Seated on the chipper truck are Chris Porter, Travis McElwee, Cody Hollenshead, and Nate Malm.

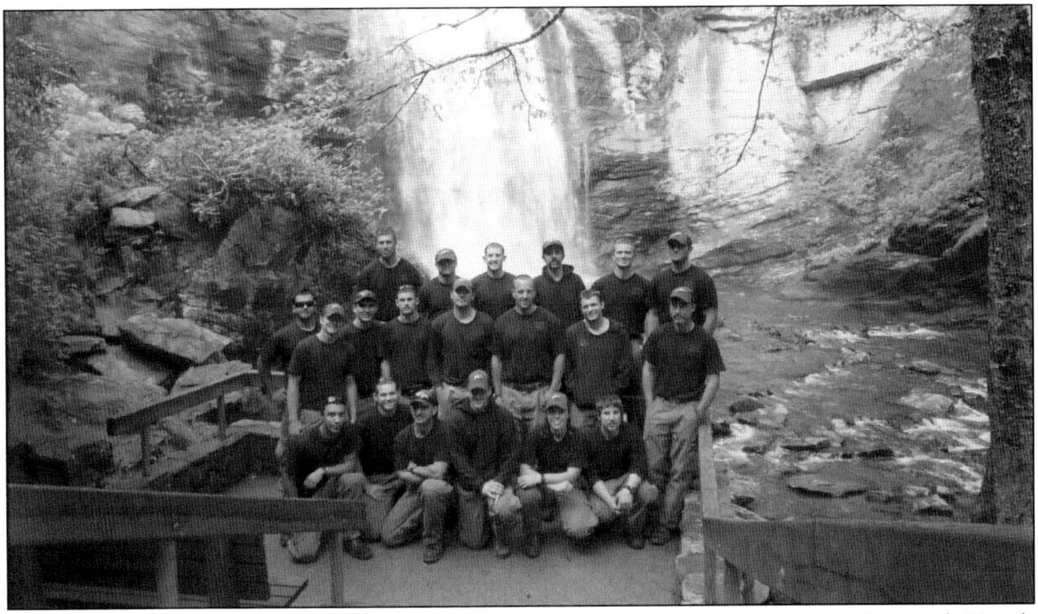

These two photographs are of the hotshot crew while on an assignment in Bervard, North Carolina. At the waterfall are, from left to right, (first row) Nick Unterman, Austin Mork, Clayton Whitted, Jeremy Sarge, Andrea Sponselor, and Matt Annibale; (second row) Ben Nelson, P.J. Fischer, Matt Barton, Dan "Boon" McCarty, Jason Bennet, Jesse Steed, Bill Kane, and Eric Marsh; (third row) Ryan Murphy, Aaron Lawson, Pat McCarty, Phillip Maldonado, Damian Sauer, and Travis Carter. Seen below from left to right are (first row) Nick Unterman and Matt Annibale; (second row) Travis Carter, Philip Maldonado, Ben Nelson, Pat McCarty, Jesse Steed, Damian Sauer, and Ryan Murphy. As is common practice among hotshot crews, experienced "fillers" are used when the crew is short of personnel. In this case, suppression firefighter Murphy is the filler. Murphy served on both a Prescott National Forest engine and the Prescott Hotshots team before coming to Prescott Fire Department. (Both, courtesy of Pat McCarty.)

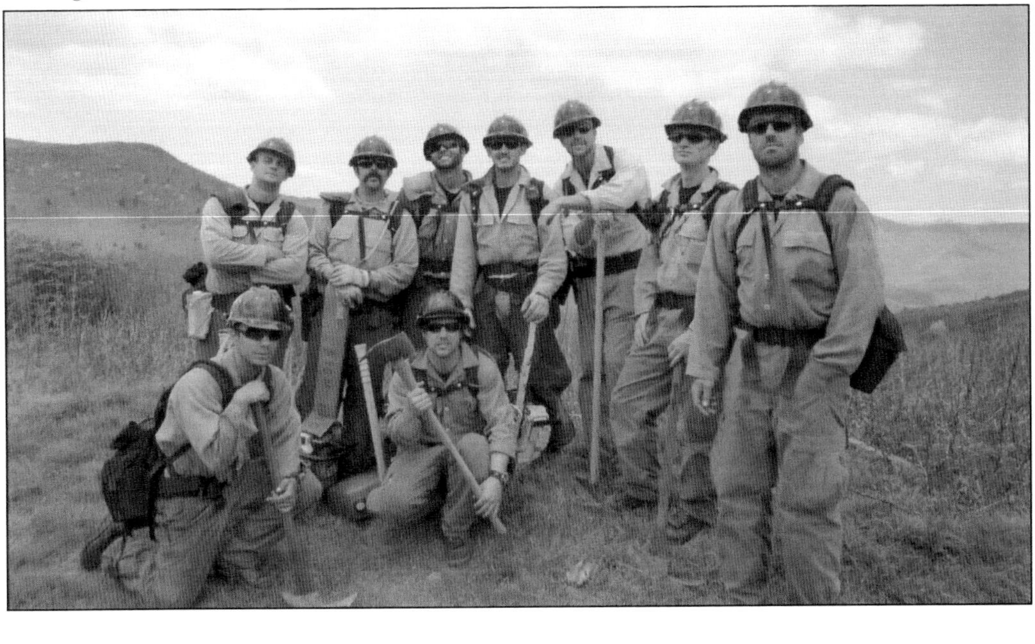

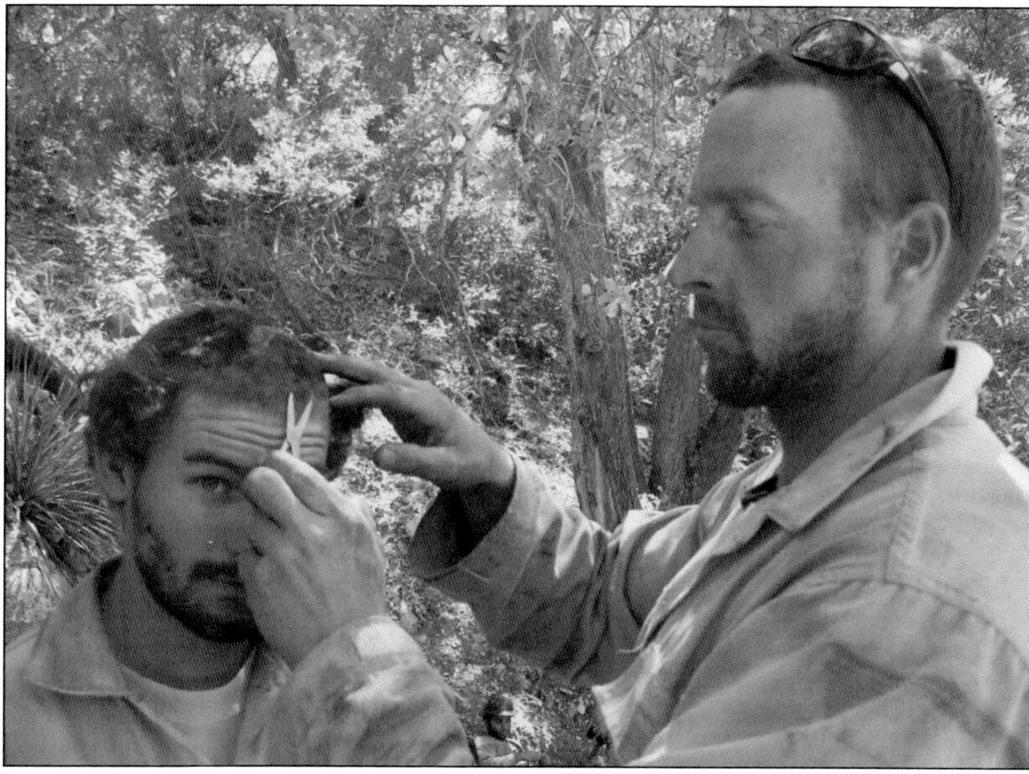

Jesse Steed gives Dan "Boon" McCarty a haircut using the miniature scissors from his Leatherman multipurpose tool. Crews can expect to be on assignments for a minimum of two weeks plus travel days to and from the fire. Given a general lack of amenities, crews often make do with what they have on hand. (Courtesy of Pat McCarty)

In celebration of Boon's 21st birthday, the Granite Mountain Hotshots pose for the camera while on assignment in North Carolina. Pictured kneeling are Jeremy Sarge and Clayton Whitted. Dan "Boon" McCarty is carried by, from left to right, Austin Mork, Bill Kane, P.J. Fischer, Matt Barton, and Jason Bennett. (Courtesy of Pat McCarty.)

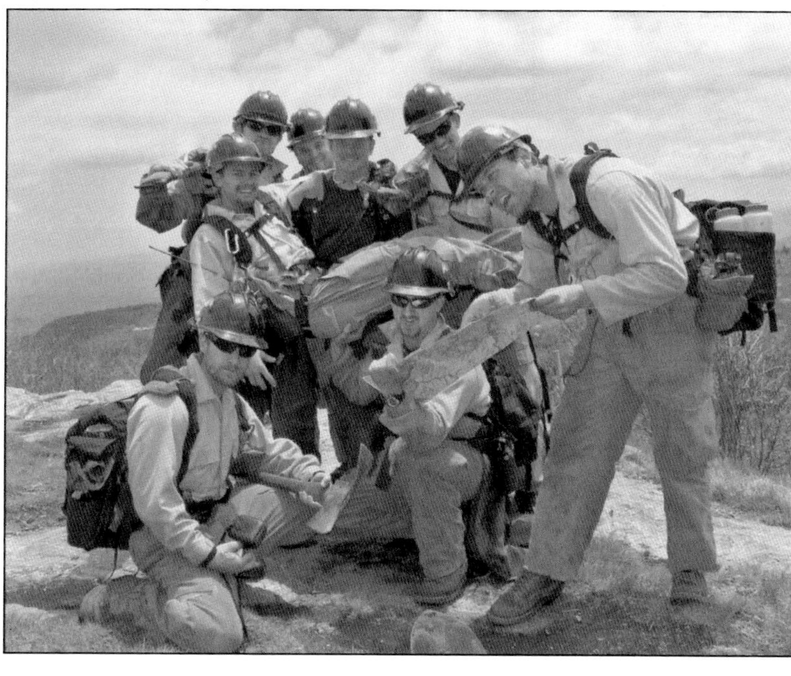

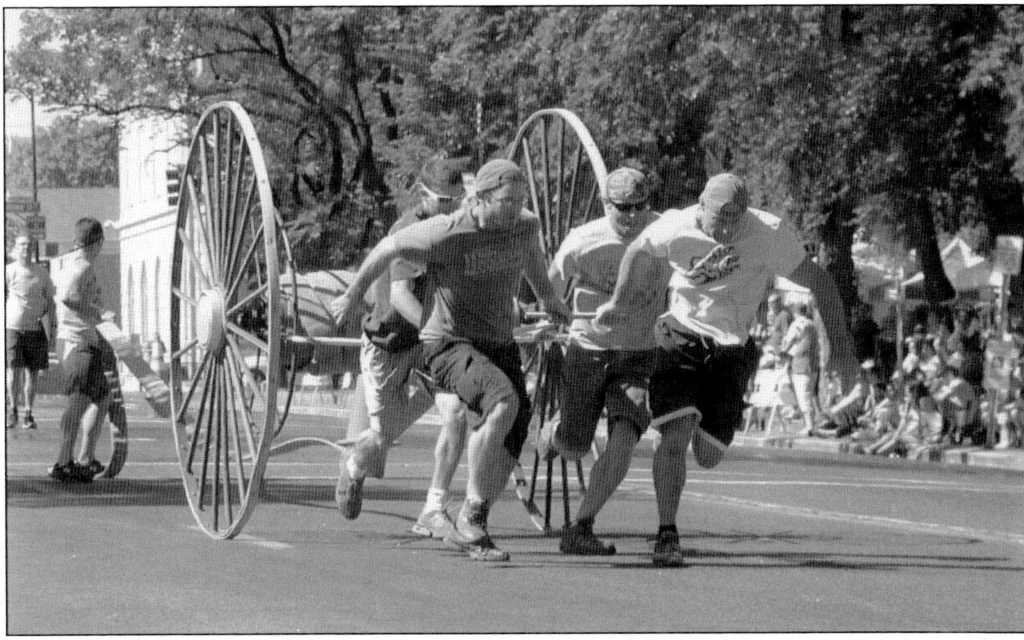

Members of the Granite Mountain crew participate in the 2010 hose-cart races. Pictured are Michael Vicente at the nozzle with, from left to right, Dan "Boon" McCarty, Austin Mork, Josh Waderski, and Jesse Steed pulling the hose cart. The crew was rarely able to attend the event because they were often dispatched to fires in other communities during the annual races.

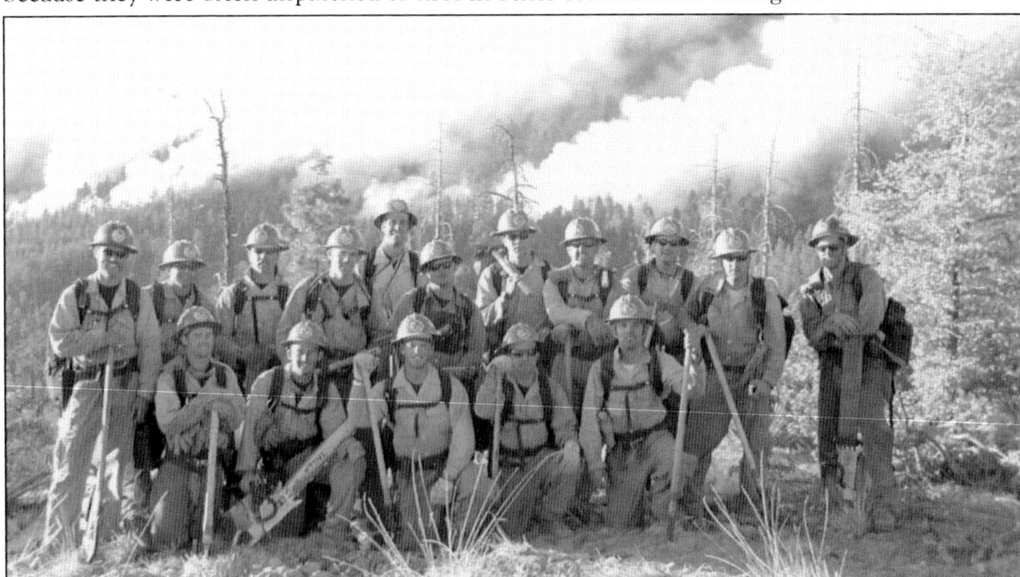

Seen here is a crew shot from the Promontory Fire in Payson, Arizona, in 2007. The image shows, from left to right, (first row) Dan "Boon" McCarty, Pat McCarty, Ted Ralston, Matt Annibale, and Phillip Maldonado; (second row) Eric Marsh, Ben Nelson, Mark Matthews, Sean Kauffman, Doug Harwood, Nate Seets, Jake Tatum, Wade Ward, Joe Hernandez, Jeremy Sarge, and Matt Bainter. Sarge, Seets, Harwood, and Pat McCarty have since taken positions in the suppression division while Ralston works as a supervisor in the wildland division.

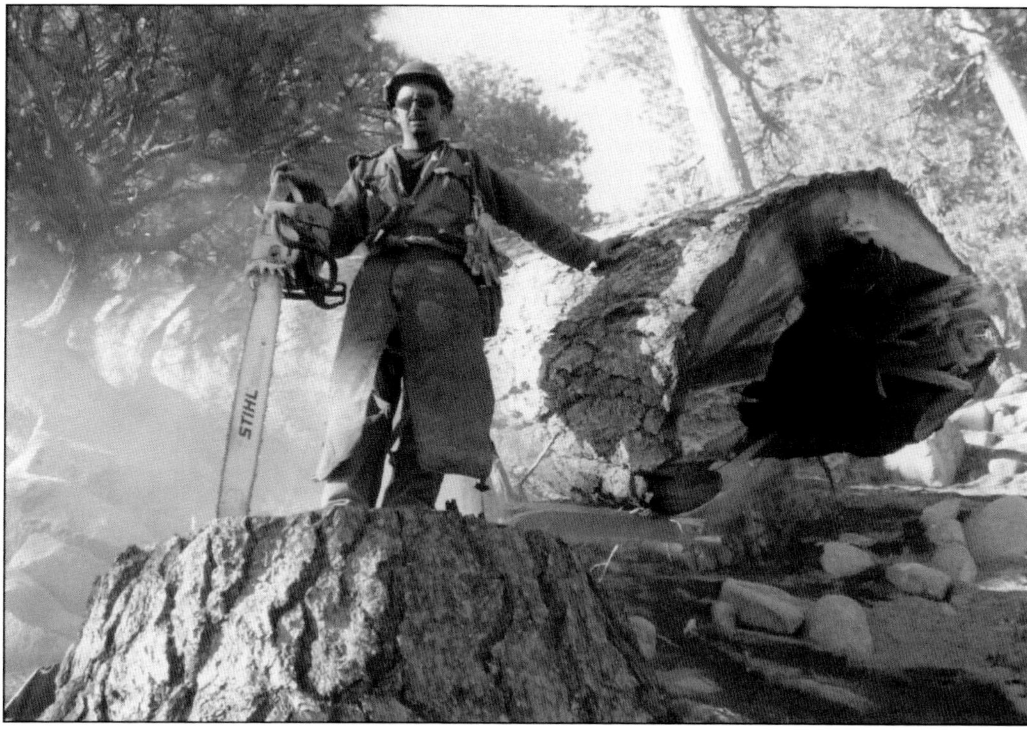

Clayton Whitted stands triumphantly over a tree he has dropped. Whitted achieved a Class C faller rating, the highest certification seen among hotshot crews, and was a saws instructor. Granite Mountain Hotshots had three dedicated saw teams, with additional backup teams for times of heavy saw work. Like most hotshot crews, all members of the crew were trained in saw use. (Courtesy of Pat McCarty.)

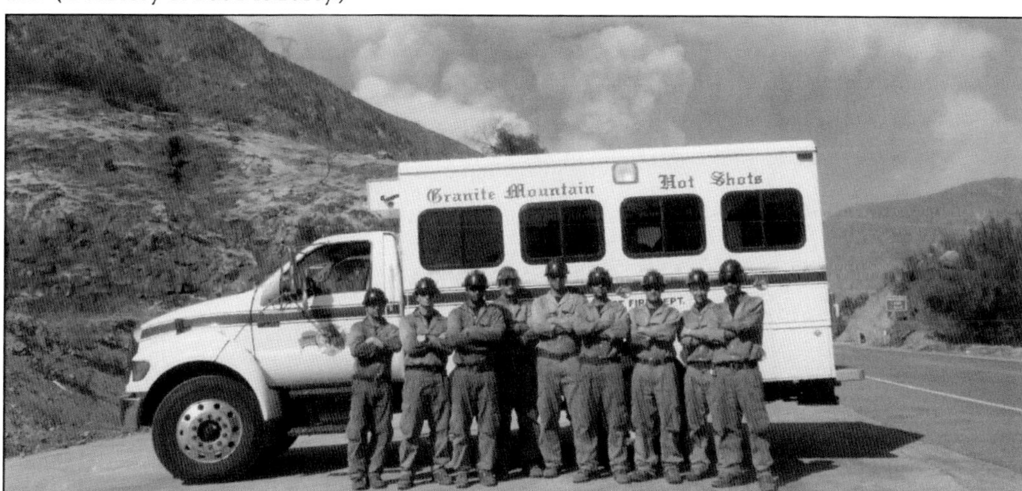

From left to right, Matt Annibale, Damian Sauer, Ben Nelson, Travis Carter, Jesse Steed, Phillip Maldonado, Pat McCarty, Josh Thompson, and Nick Unterman pose in front of their crew buggy at the Station Fire in 2009 in Southern California. The crew initially operated out of passenger vans before it purchased crew carriers. While the federal hotshot crews use green carriers, the Granite Mountain crew chose white. (Courtesy of Pat McCarty.)

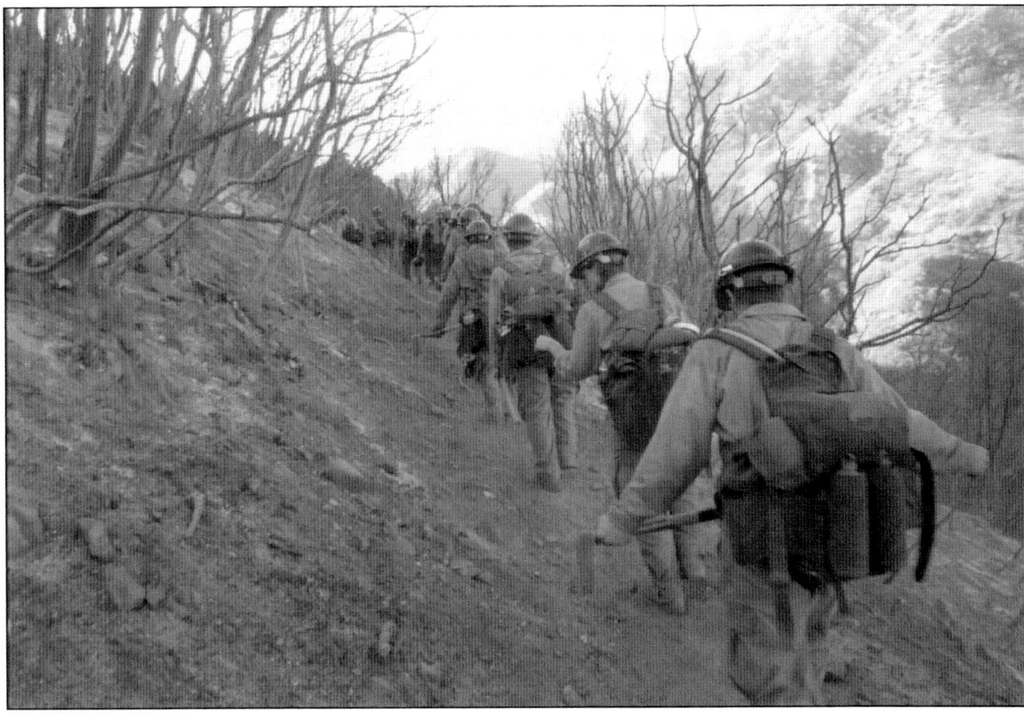

The Granite Mountain crew marches in a line through the black of the Station Fire in 2009 to get to their assignment. "Lining it out" is standard practice among hotshot crews, maintaining the unit as a whole and establishing the work order when the crew constructs fire line. Constructing hand line around a fire is slow, tedious work and each member of the line only makes a small impact in the construction. The whole of each member's contribution leads to miles of construction over the course of their 16-hour shifts. (Both, courtesy of Pat McCarty.)

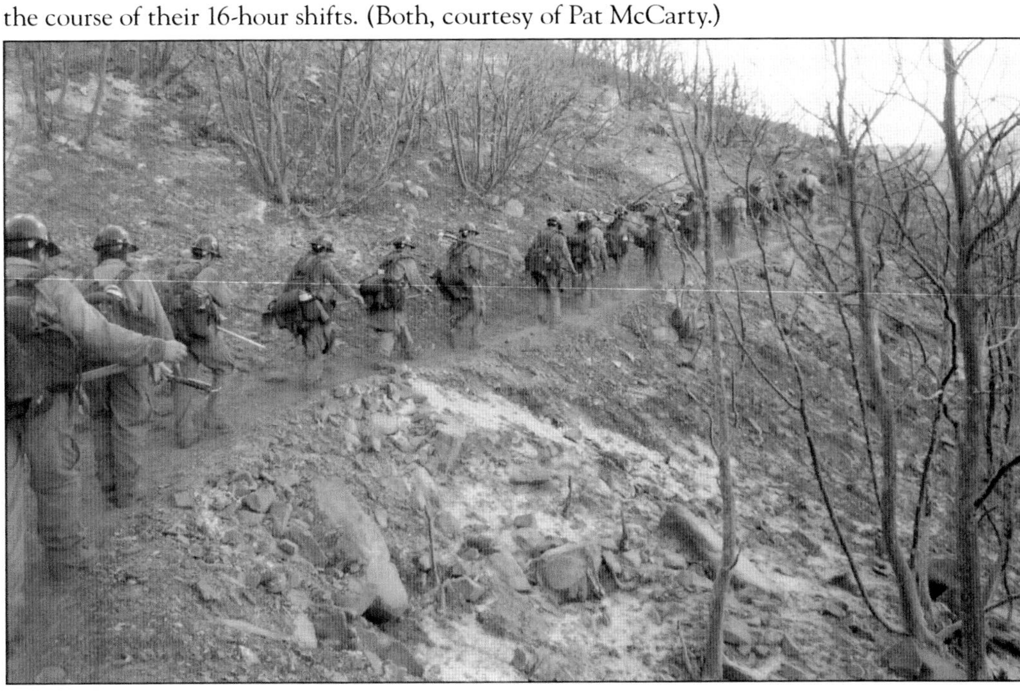

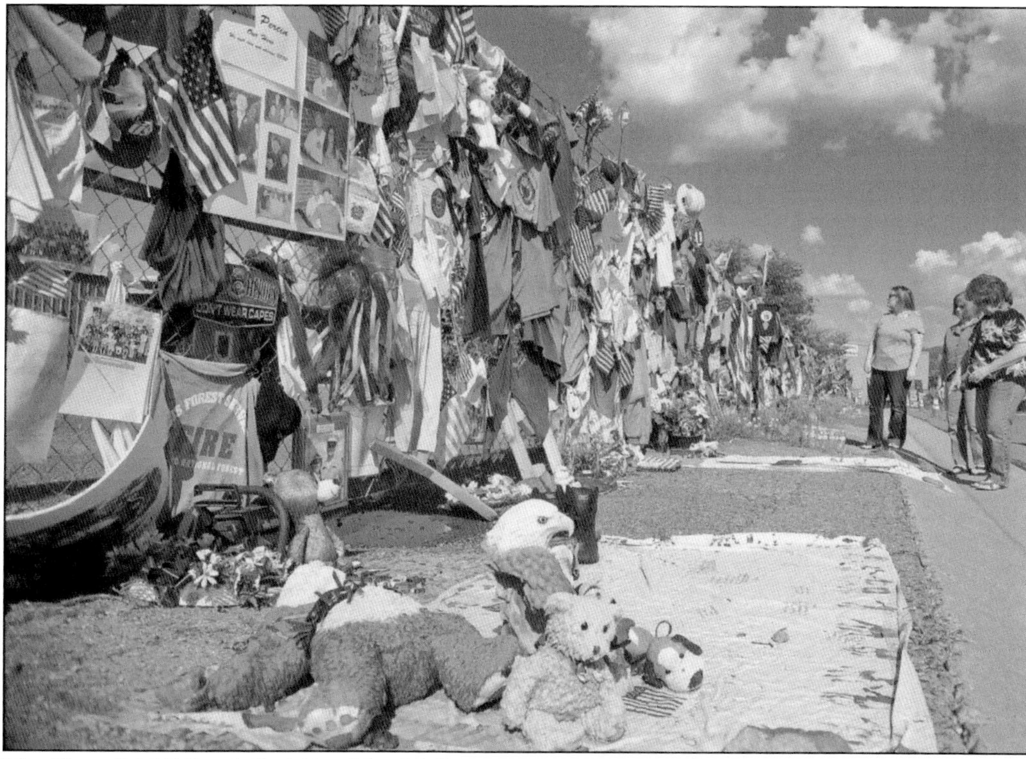

The Yarnell Hill Fire killed 19 of the 20 Granite Mountain Hotshots on June 30, 2013, when the winds from a thunderstorm pushed the fire into the heavy brush surrounding their position west of Yarnell. The chain-link fence surrounding Granite Mountain's Station 7 became a monument to the lost men within hours of their passing, with people traveling thousands of miles to pay their respects. At the time of the disassembly of the memorial fence on September 10, 2013, some 1,100 shirts were removed, along with 900 American flags, 496 stuffed animals, 851 prayer flags, and other memorabilia. Community volunteers catalogued and preserved every memento from the fence for future display. Each section of the fence can be recreated exactly as it stood on September 10, 2013. (Above, courtesy of Matt Hinshaw; below, courtesy of Wade Ward.)

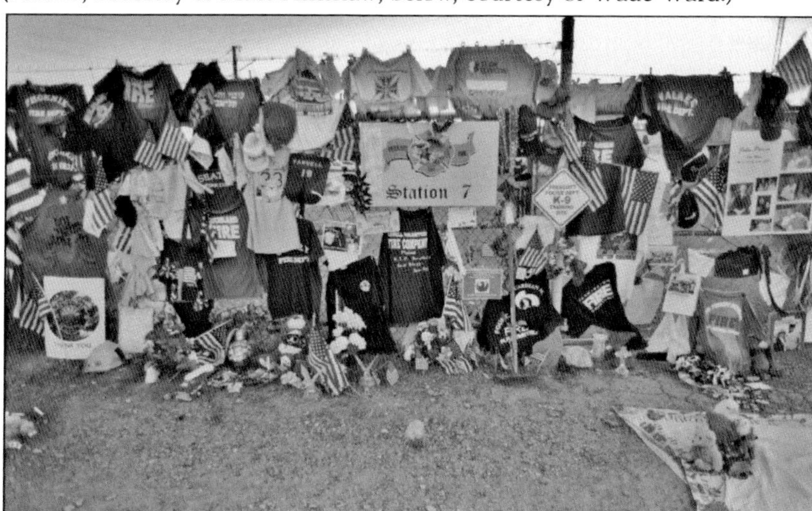

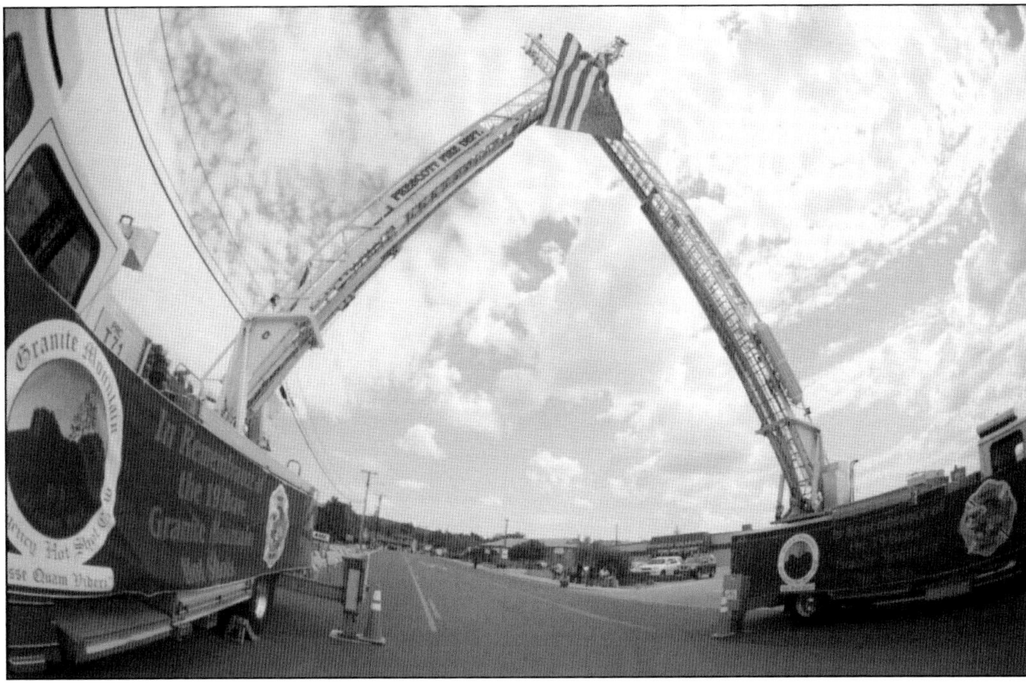

The bodies of the fallen 19 Granite Mountain Hotshots were returned to Prescott on July 7, 2013. At the request of their families and the Prescott Fire Department, their route passed through Yarnell, where they lost their lives, and then passed their crew quarters in a symbolic return to their station before being taken to the Yavapai County Medical Examiner's office to lay in wait. (Above, courtesy of Wade War; below, courtesy of Matt Hinshaw.)

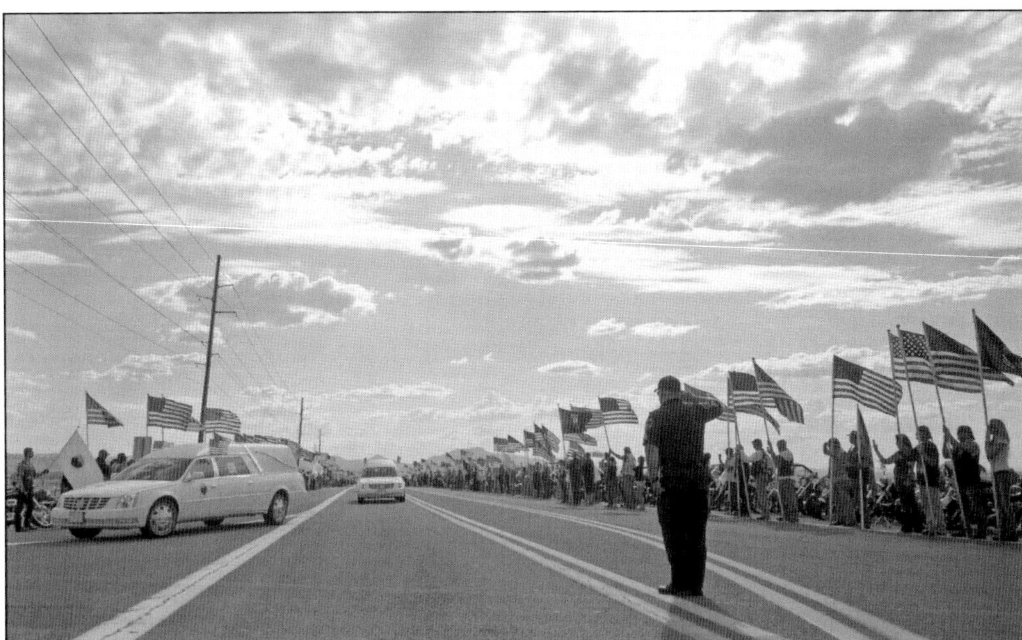

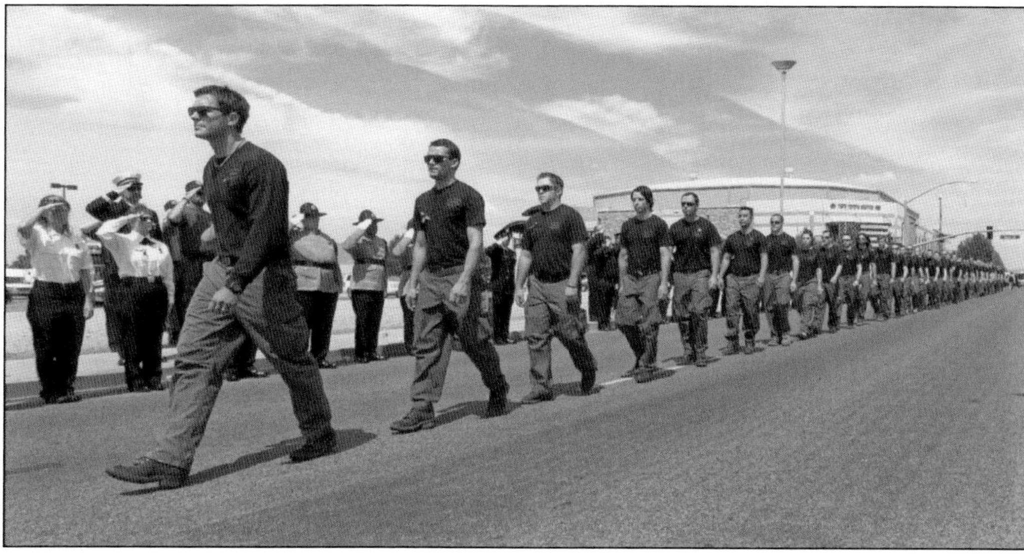

Almost 50 alumni from the Granite Mountain Hotshots march in their traditional line on the way to the memorial ceremony at Tim's Toyota Center. The honor of being alumni of this hotshot crew belongs to 67 men and women. Some 20 percent of Prescott Fire Department's suppression roster is composed of men who have served on Granite Mountain Hotshots, either as firefighters or fillers on the crew. Representatives from all 108 hotshot crews, innumerable fire department honor guards, Arizona governor Jan Brewer, and Vice Pres. Joe Biden attended the memorial. (Both, courtesy of Matt Hinshaw.)

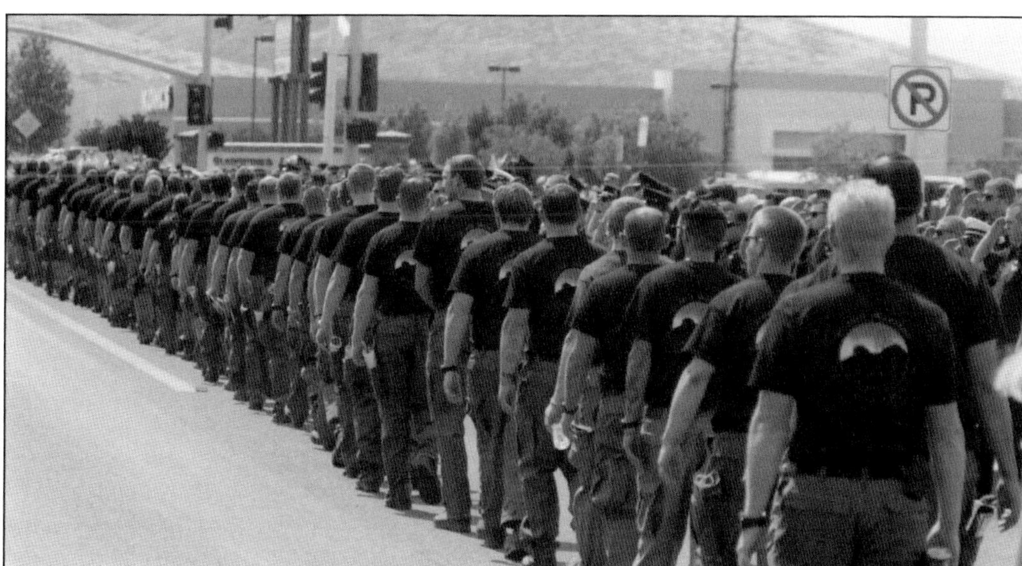

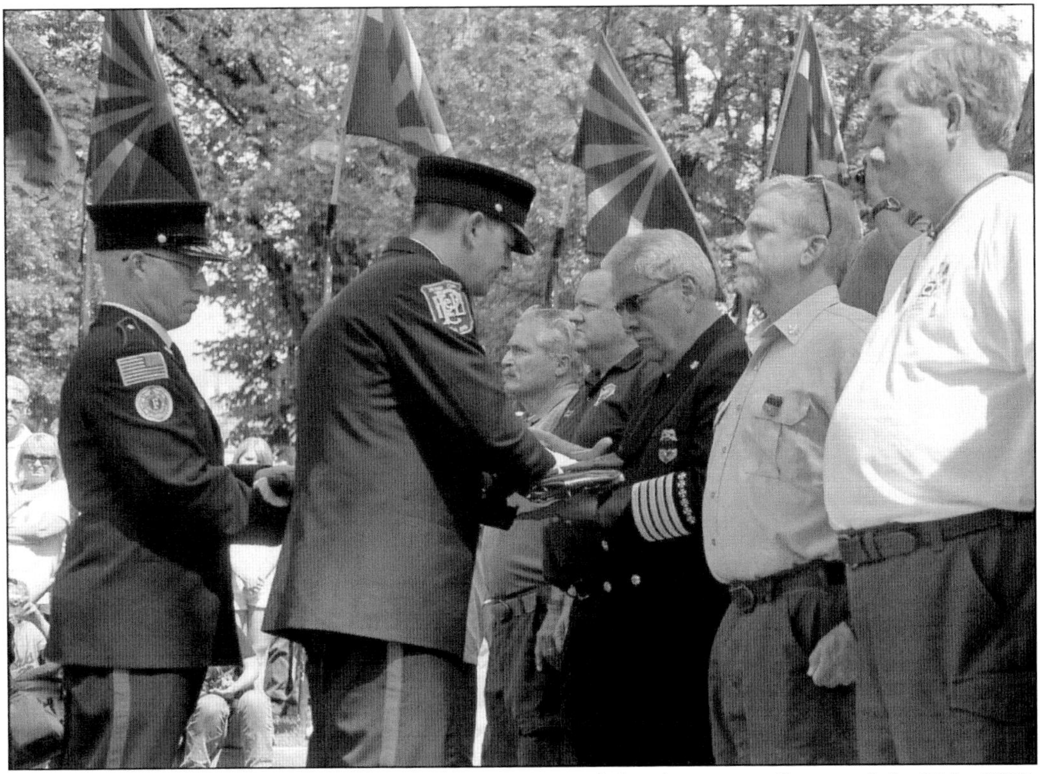

Prescott Fire Department chief Dan Fraijo is presented the American flag at a July 13th, 2013, ceremony at the county courthouse. From left to right on the receiving side are Tony Sciacca, Eric Kriwer, Dan Fraijo, Pete Gordon, and Mike Quisenberry. The ceremony was a symbolic and formal closure of the honor guards' protection of the hotshots' bodies, which they had protected since July 1, when Prescott firefighters brought them to the medical examiner's office. (Courtesy of Matt Hinshaw.)

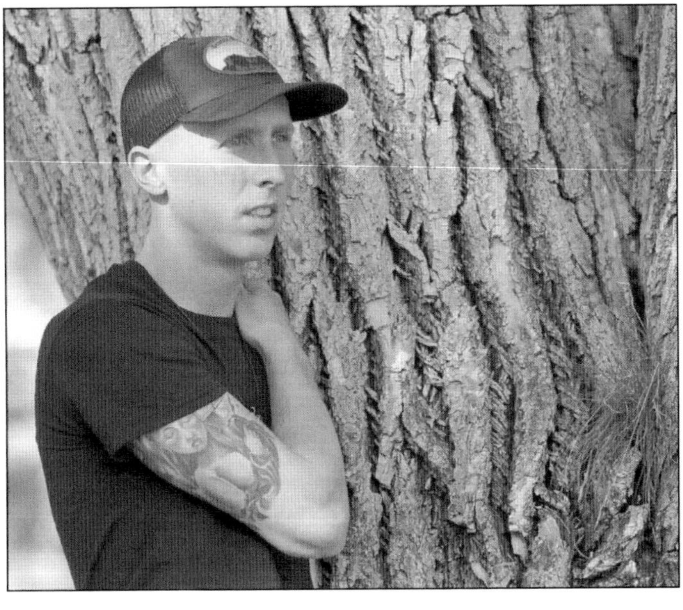

Brendan "Donut" McDonough was Granite Mountain Hotshots' sole survivor from the Yarnell Hill Fire. McDonough was in his third season and served as lookout for the crew on June 30, 2013, prior to relocating his position when the fire hit his trigger point for his own evacuation and relocation. The Blue Ridge Hotshots gave McDonough a ride back to his crew carriers, effectively removing him from the hazard area. (Courtesy of Matt Hinshaw.)

Eight

Excellence in Service

Prescott Fire Department's history extends past its formal establishment in 1885 to the year 1875, when the city first started purchasing ladders and hooks. In the nearly 140 years since, the department's firefighters have met the city's needs with a standing service army, which has had to diversify over the century. During that time, their involvement in the community has extended beyond their firefighting and medical roles to include a wide array of other social and service aspects.

Firefighters are widely regarded as one of the most trusted professions, with citizens having faith in firefighters to protect their health, lives, and most beloved possessions during both fire and medical events. Prescott's firefighters have met the expectation and refer to it as "excellence in service," but that service extends beyond the expectations to follow an above-and-beyond mindset. As is often joked about in the firehouse, firefighters are all things to all people, rising to the needs of the situation, whether mundane or extraordinary, for the wealthy or destitute, in weather pleasant or foul, and at any hour.

A requisite component in being able to meet those needs is an innate understanding of and interest in their community. As such, Prescott's firefighters are involved in the things that make this community one of the best in the nation. They participate in old-fashioned parades, partake in community service projects, and honor their traditions. In 2013, they mourned their fallen brothers publicly along with their community, drawing them together in a way that few other cities in the nation have experienced. The firefighters were able to experience a community that supports them in thanks for the service they have rendered for 140 years.

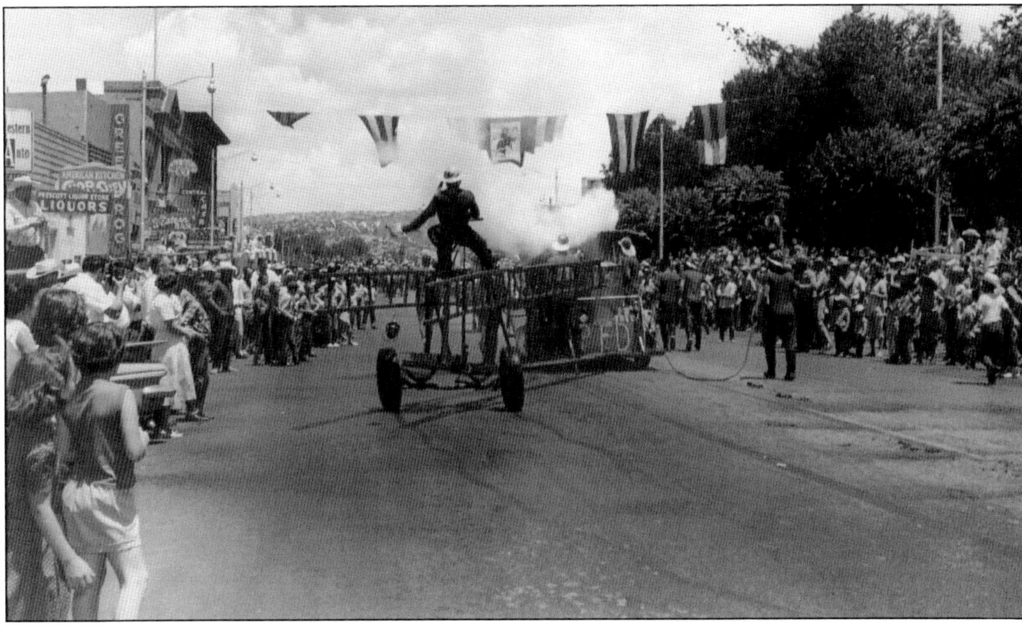

The infamous "Custer's Last Stand" entry in the 1957 Frontier Days parade took the fire department entries to a new level. Where the tradition had been simply the hose carts, wagons, and engines of the era, the firefighters this year fabricated a pseudo tiller truck from a 1952 Hudson that included an operational tiller component and a functioning hose line that could be used to spray the crowds. Chief Custer, who had only been in charge for one year, would have the final word in regards to the tongue-in-cheek name of the entry. He ambushed the entry at the corner of Montezuma and Gurley Streets with a hose line attached to a hydrant. (Above, photograph by Matt Culley.)

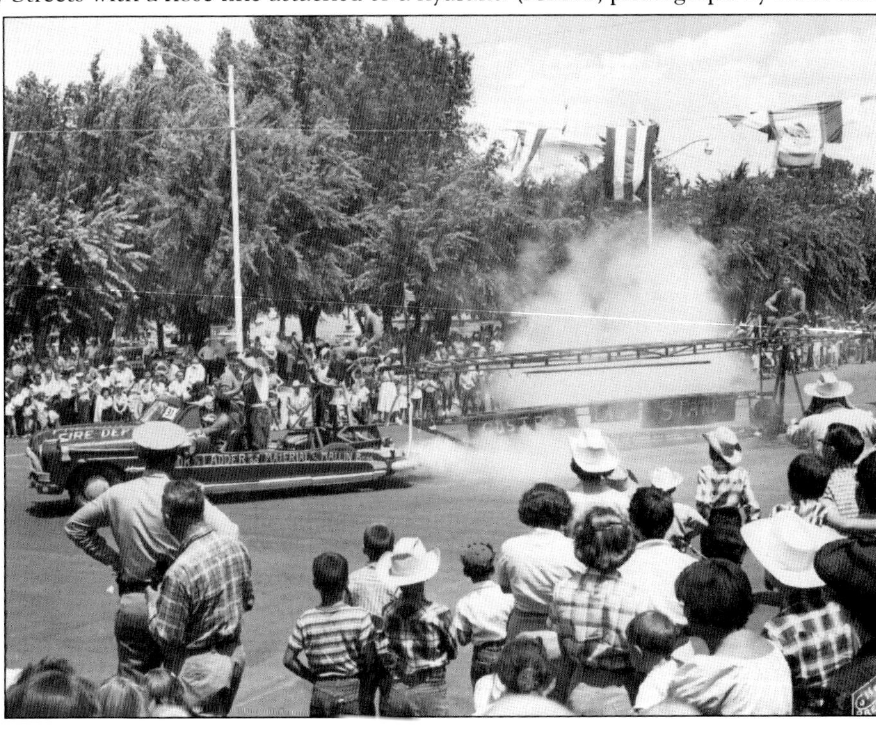

The firefighters' 1958 Frontier Days parade entry won in the best all-around category. Firefighters cut and re-welded the tire rims to be off-center, causing the vehicle to wobble down the street in a haphazard fashion. The firefighters continued the tradition of wearing flannel underwear in the event.

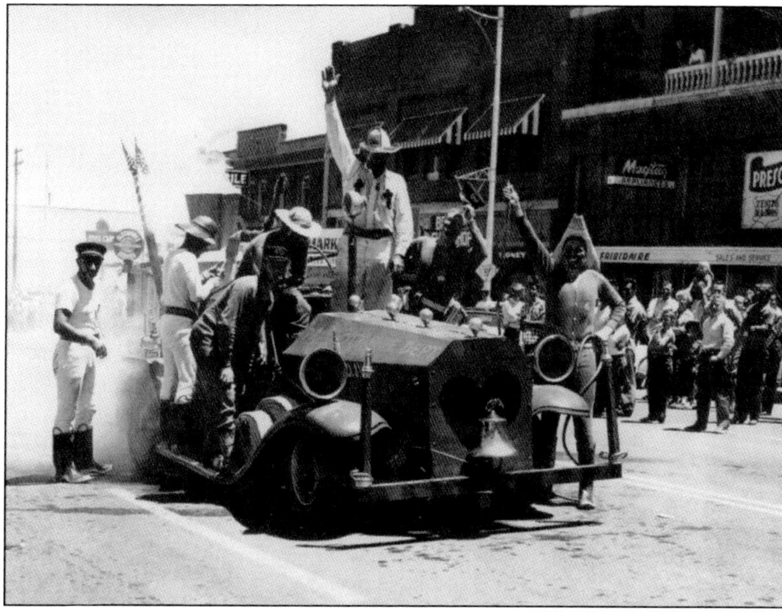

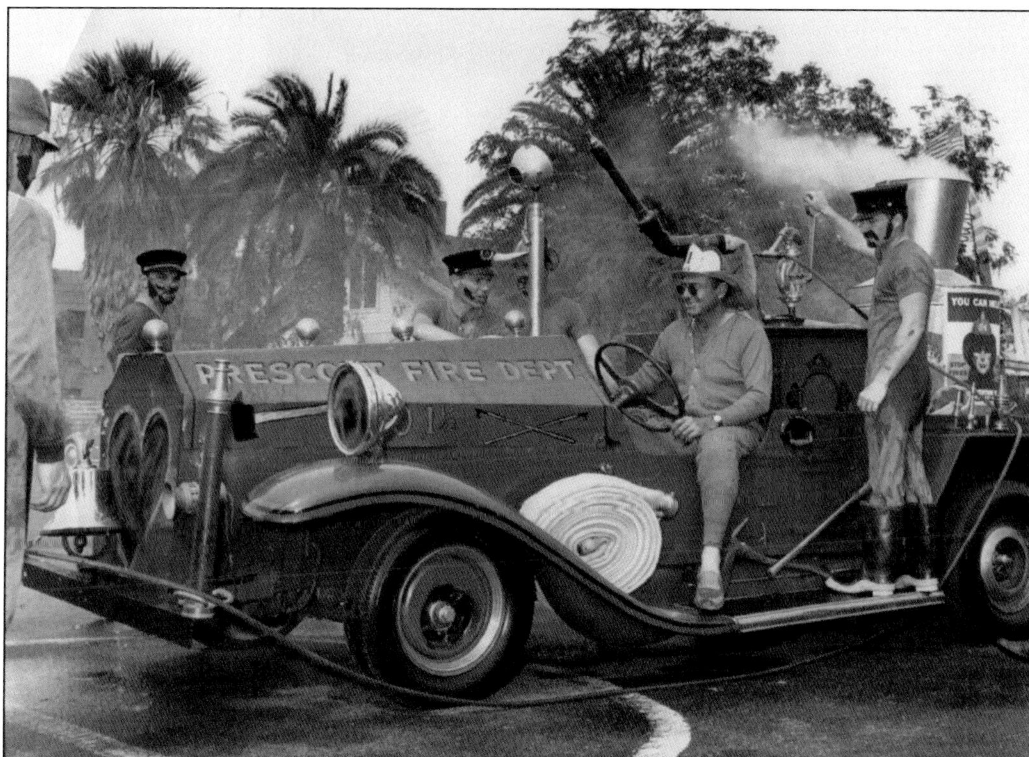

As a part of Fire Prevention Week in 1958, Prescott firefighters were invited to drive their award-winning entry in the Phoenix Fire Department Prevention Parade. The Prescott firefighters ended up taking home the grand prize. Seen here from left to right are unidentified, Vic Lira, unidentified, Josh Spivey, Sid Webb, and George Ravetto. (Photograph by Phoenix Chamber of Commerce.)

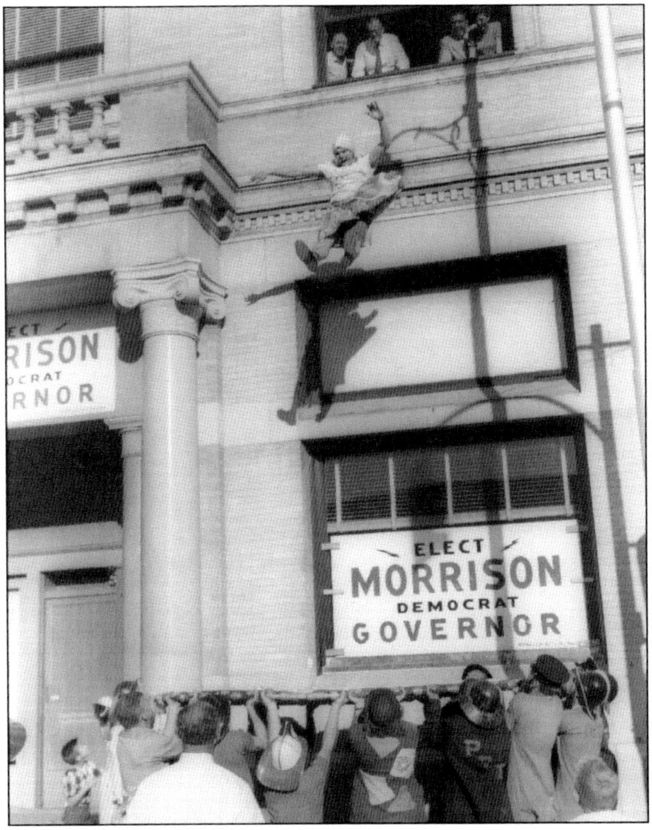

A rite of passage among the firefighters of the time was jumping off the firehouse on Goodwin Street to the awaiting rescue ring below. Here, Gene Shimonowsky takes it a step further, jumping from the second story of a downtown building. Shimonowsky was known for his willingness to jump to the rescue ring being held below. Unfortunately, this would be his last jump. His fall was inadequately broken and he struck the concrete sidewalk hard enough to injure his back. The practice of jumping from the building was discontinued after firefighters received injuries from the jumps. The jumpers were not the only ones being hurt. The holders of the ring had to hold it in a very specific manner in order to absorb the impact of the falling person. Failure to do so resulted in hand, wrist, and rib injuries.

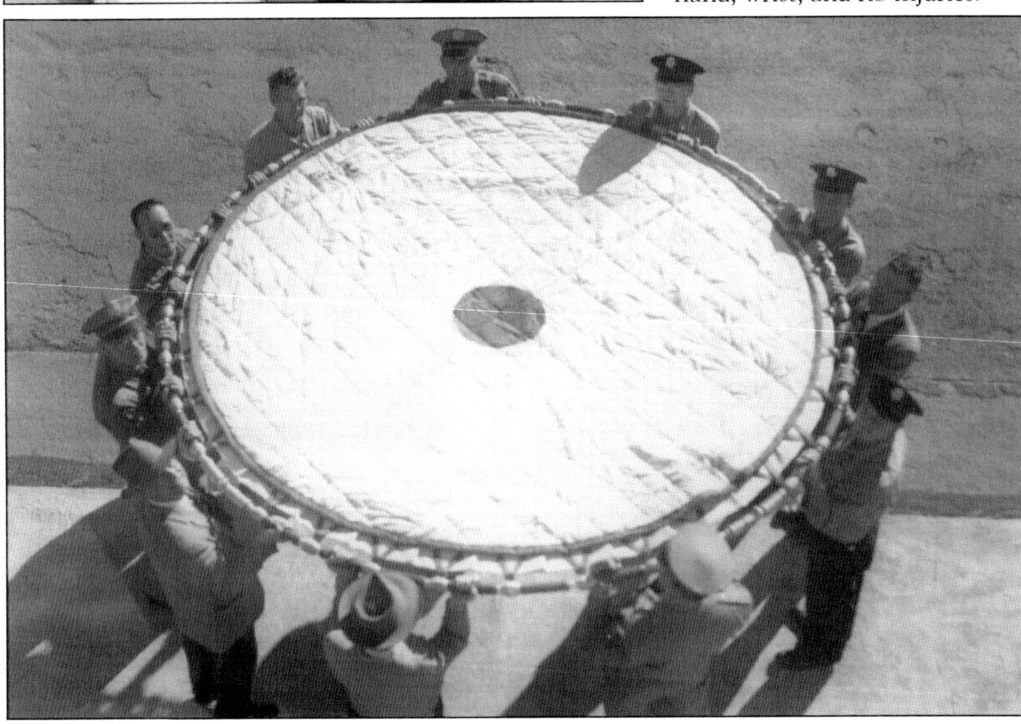

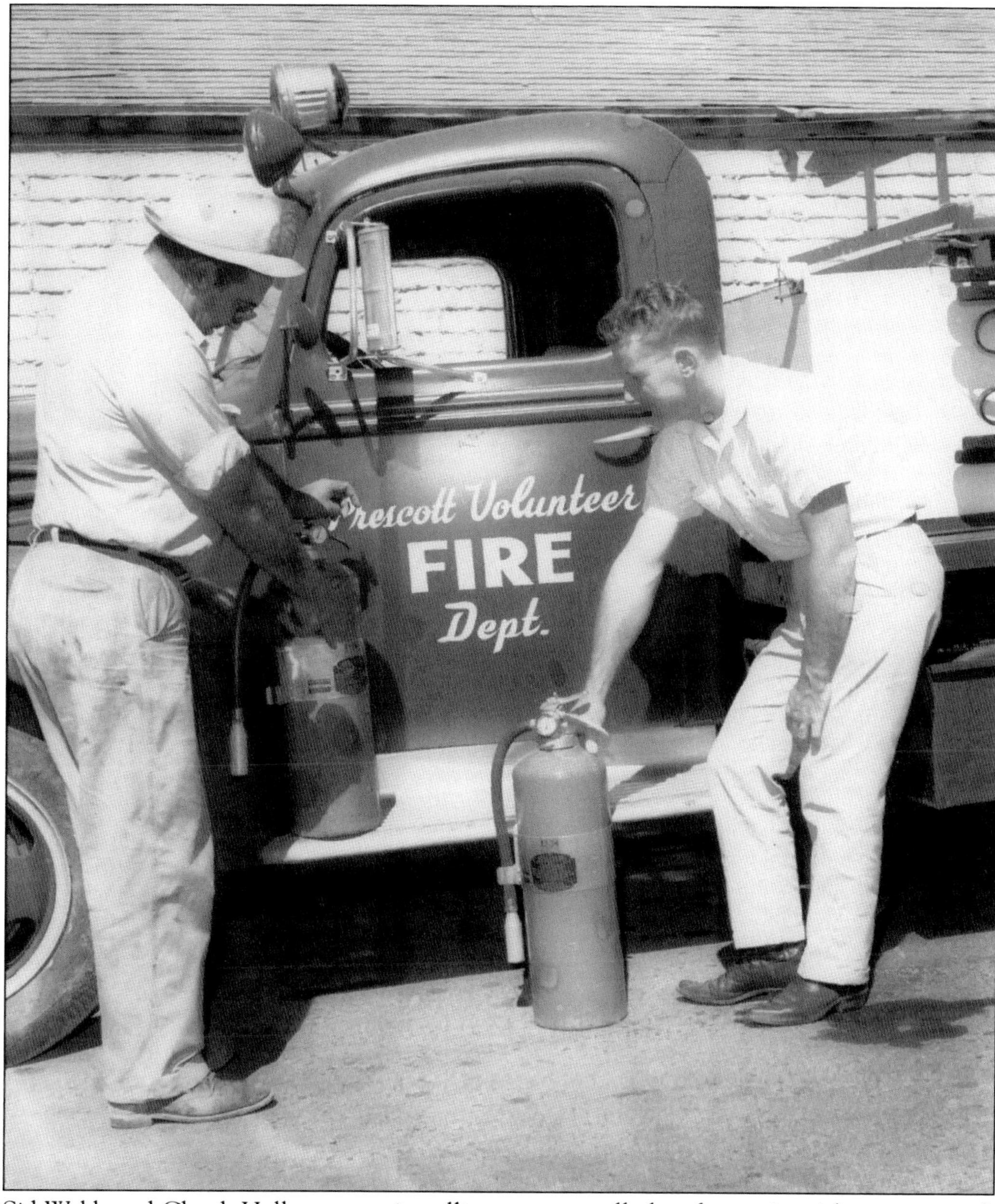

Sid Webb and Chuck Halleman are installing two new all-class fire extinguishers on a truck that would be used to fight fires outside of the city limits. The truck was donated by Hassayampa Mountain Club in 1960 and housed in a building behind the Goodwin station. Ten of the volunteers signed up each month to respond to fires occurring outside the city limits. Donations of equipment for the engine poured in following the donation, including radios, axes, and even a paint job. The city later deemed the truck and the volunteers' efforts to be a liability to the city, and Mayor Tutt gave a directive that Prescott firefighters should not respond outside of the city. Sid Webb, president of the Prescott Volunteer Firemen's Association, responded, "They'd have to tie us all up with a rope to keep us from going."

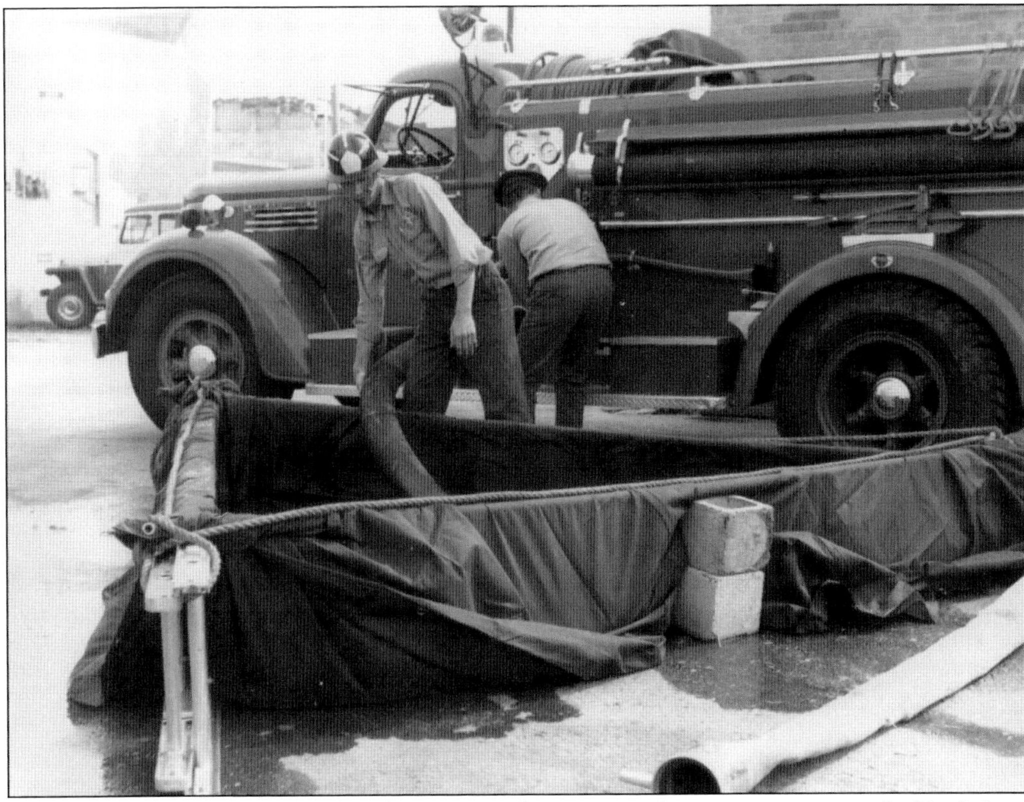

Firefighters drilling behind the Goodwin station demonstrate the ingenuity expected of firefighters in a profession that dictates creative solutions to unique problems on a regular basis. Here, the firefighters have improvised a water tank from ladders and salvage covers. The need for such a device was common in Prescott, where limited access to water supplies forced firefighters to conduct water shuttle operations. Local water tenders now carry a folding tank to accomplish this same function when fighting fires in the outlying areas.

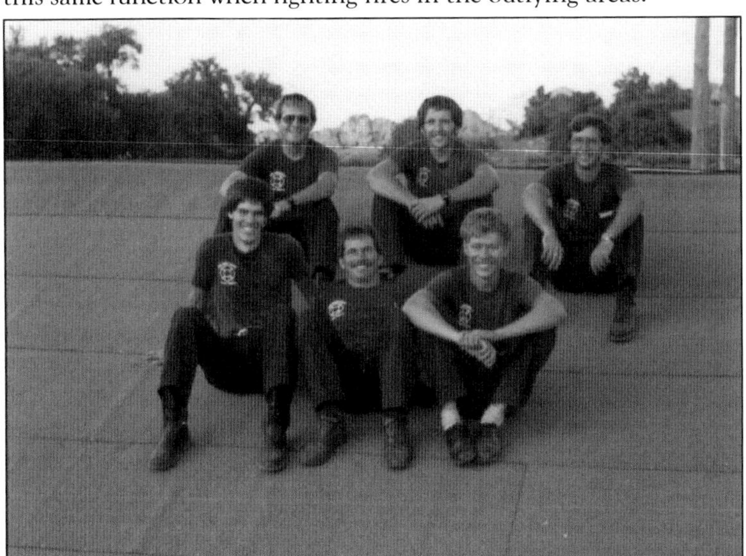

Seated on the newly constructed roof prop at the Prescott Training Center are, from left to right, (first row) Bruce Martinez, K.C. Scott, and Marty Cole; (second row) Ken Ryer, Brad Malm, and Paul Laipple. The burn tower of the training center and some outbuildings were built by the firefighters themselves for approximately $65,000, a significant savings to the outside bid of $1.5 million.

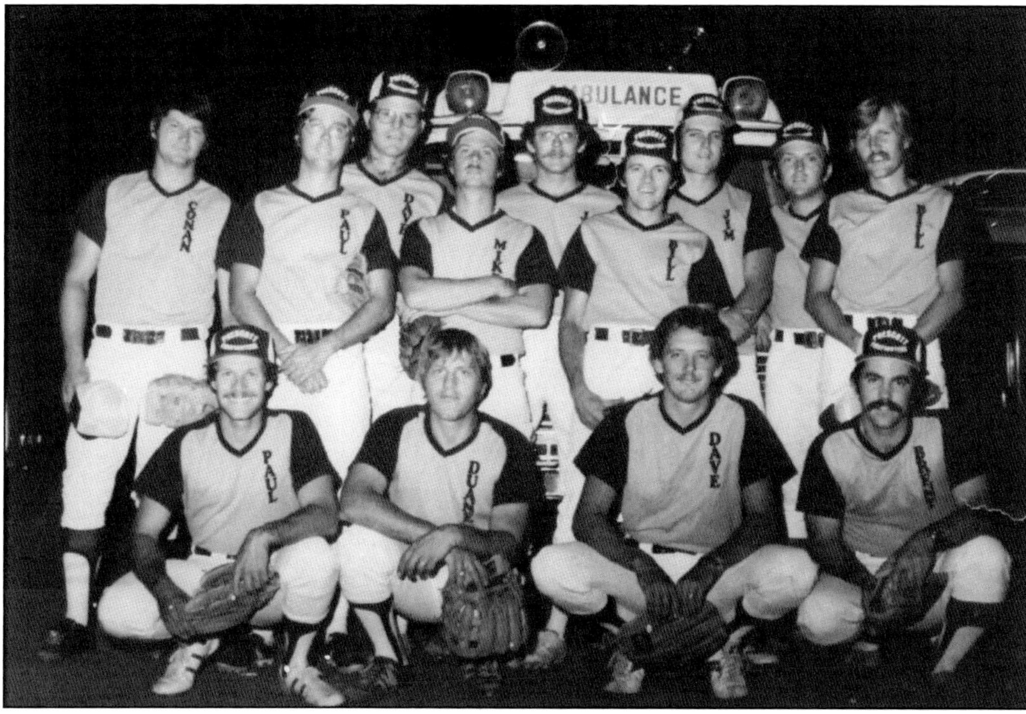

Participation in local athletic activities is a staple in the firehouse. With Prescott's national designation as a softball destination, it is natural that there have been ongoing softball teams. In this 1980s team photograph are, from left to right, (first row) Paul Coe, Duane Steinbrink, Dave Temple, and Brent Schnitzius; (second row) Conan Quick, Paul Laipple, Dave Doty, Mike Parrish, Jim Cook, Bill Wagoner, Jim Mason, Brad Malm, and Bill Heuer.

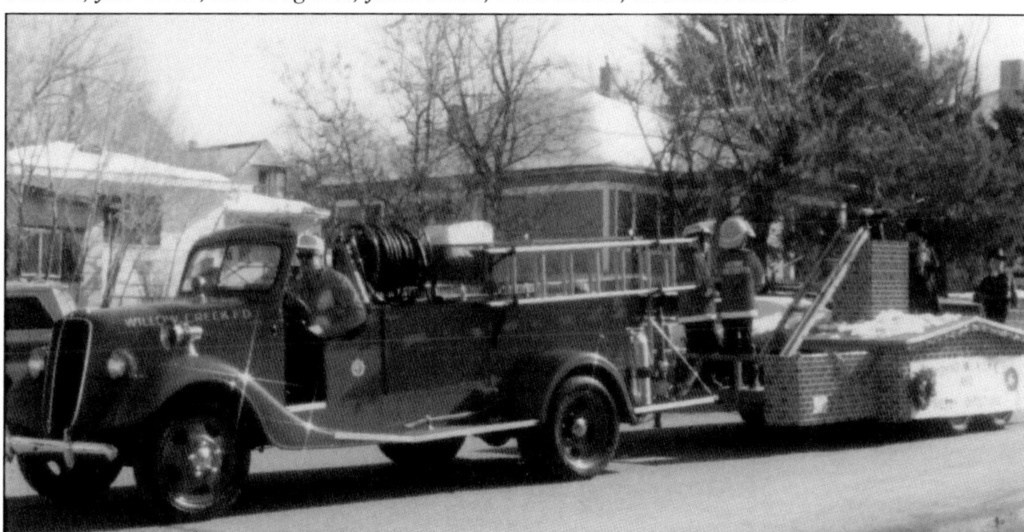

Dean Hinshaw drives the personal fire truck belonging to his father, Bob, in the annual Christmas parade. On the float and wearing "future firefighter" outfits are Dean's sons, Billy and Robby. Both boys would later take a fire science class in high school, and Robby would eventually work on the Granite Mountain Hotshots.

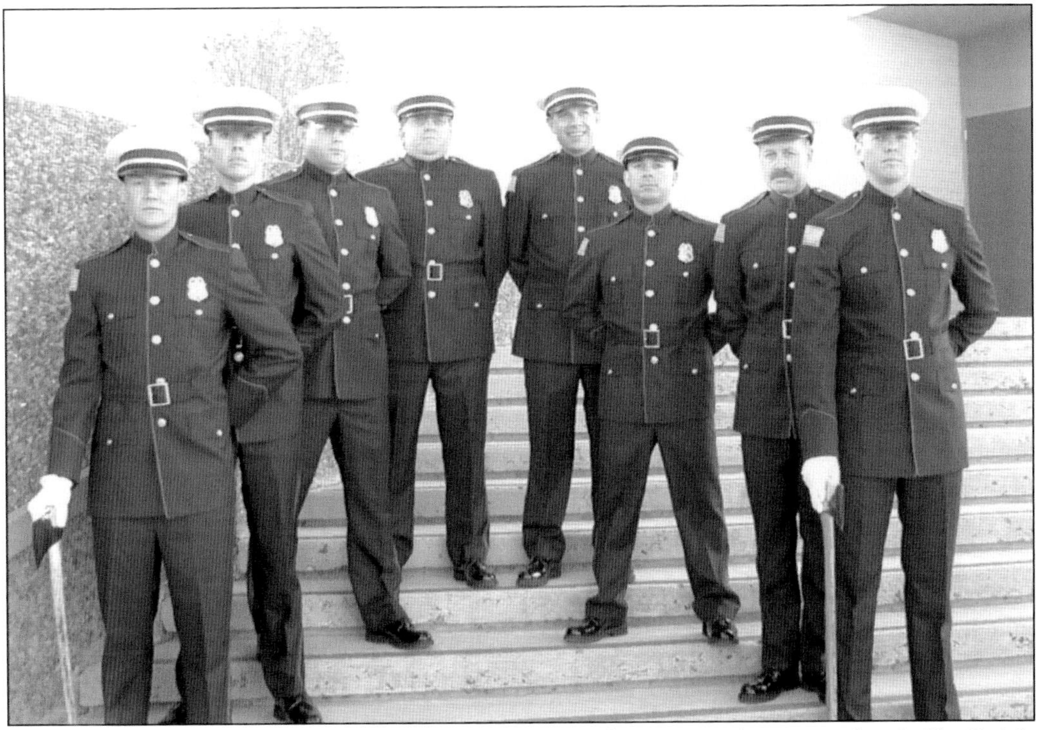

The department's honor guard was developed in the early 2000s and was spearheaded by Ralph Lucas. City council allotted $5,000 to get it established. The bulk of those funds went to the purchase of the military-style uniforms worn by Prescott Fire Guard members. Each uniform was custom fitted to the firefighter. Ultimately, the guard seeks to add bagpipes and drums to its complement. From left to right are Dan Bauman, Jeremy Brinkerhoff, Josh Groves, Brad Malm, Ralph Lucas, Dan Bates, Dean Hinshaw, and Dave Peterson.

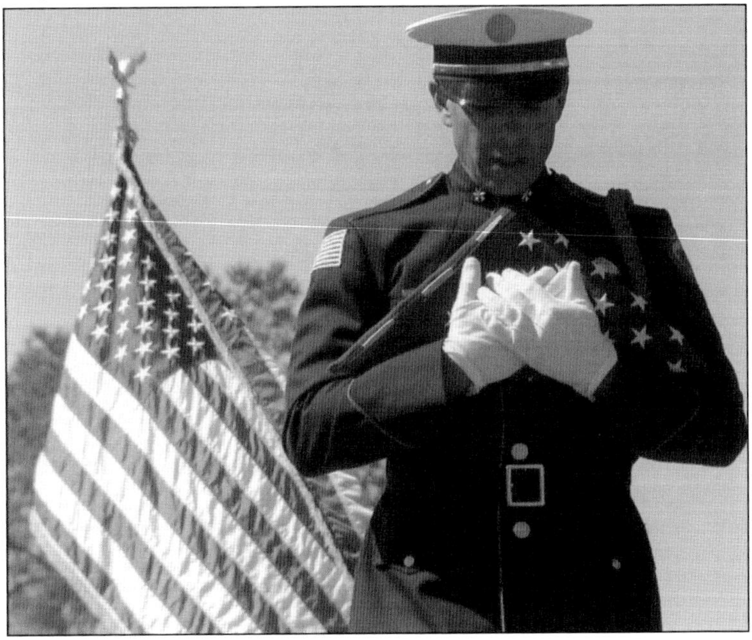

Ralph Lucas prepares to present the flag to Brad Malm's family at the honor guard's first funeral service. Malm served the department for 30 years, rose to the rank of battalion chief, and was a fire science instructor for Yavapai College. The guard's presence was required extensively in 2013 due to the loss of the Granite Mountain Hotshots.

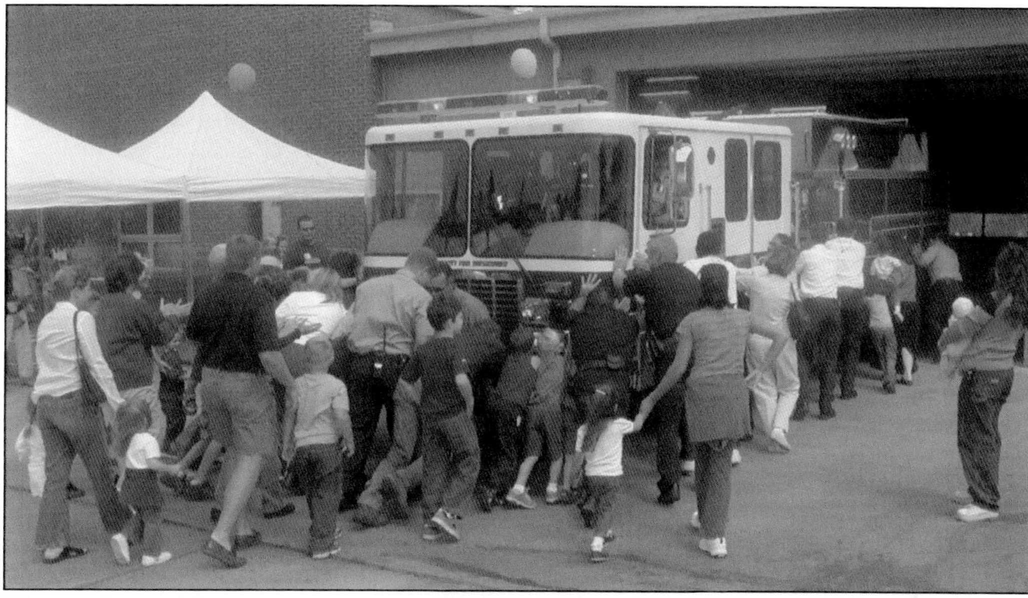

The purchase of the "Sisters," two HME 1,000-gallon-per-minute engines in 2003, saw the introduction of a traditional housing ceremony, which was established in early Philadelphia. The engine is pushed into the station three times, in a reflection of the holy trinity, by the firefighters and the citizens the engine serves. Alternately, it is sometimes stated that the number three is for the three groups the engine serves to protect: the community at large, its response district, and the firefighters themselves. Every new engine and truck company has been ceremoniously housed since the observance was introduced. (Both, courtesy of Wade Ward.)

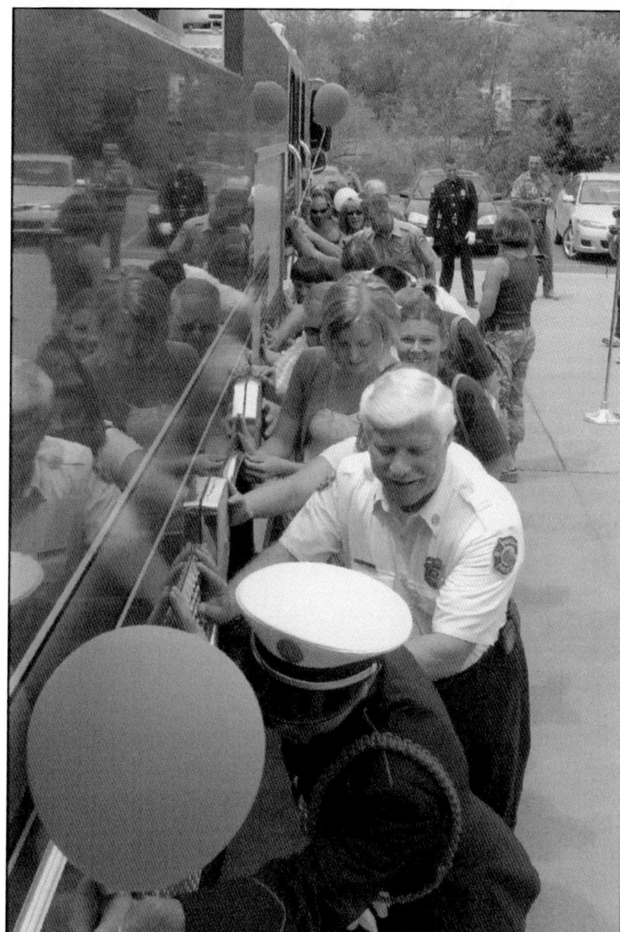

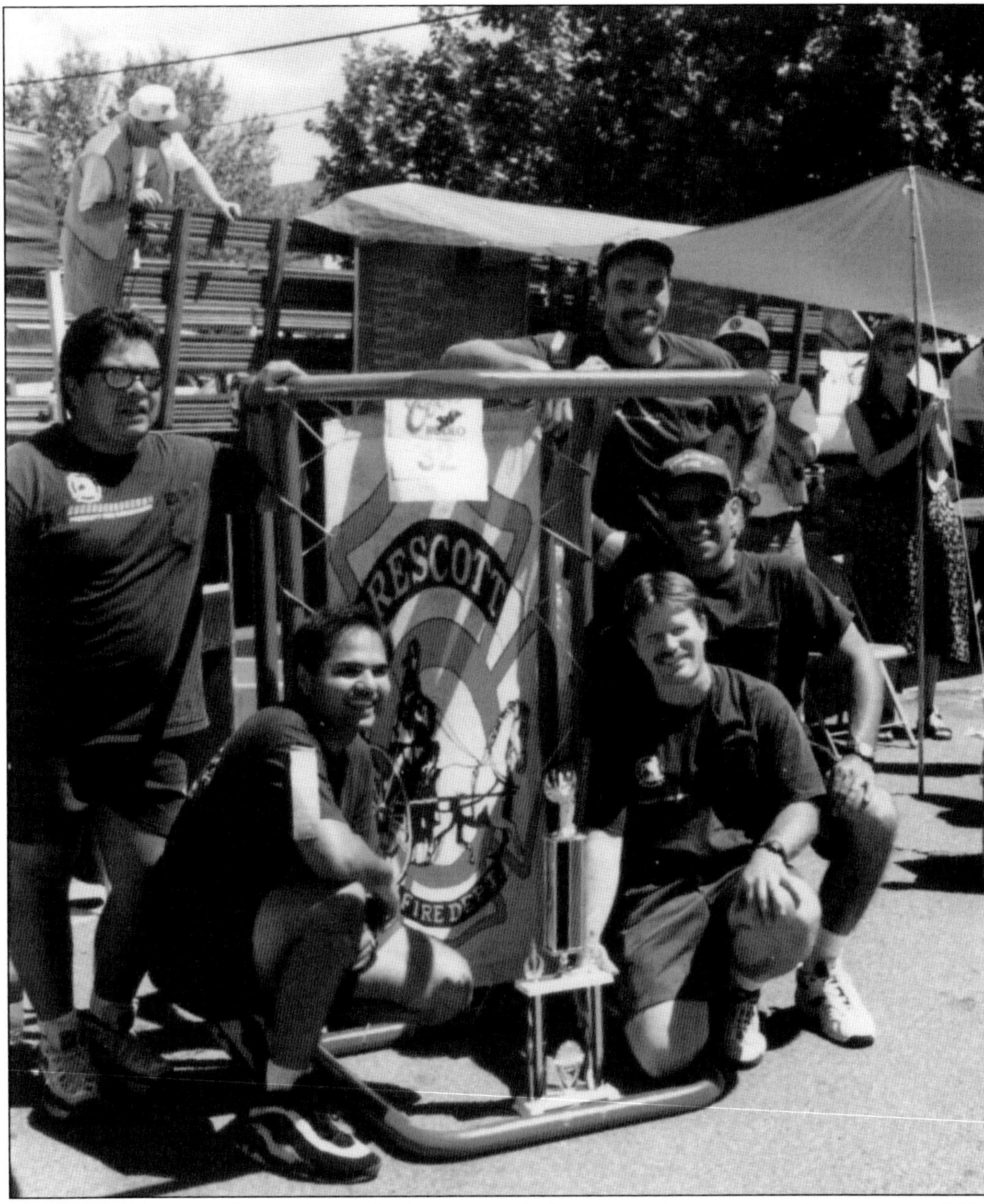

In 1997, the Prescott Sunrise Lions Club sponsored a bed race where the winning team would be sent to Honolulu, Hawaii, to compete in the international bed race competition. With a vacation on the line, the race was loaded with competitors. Prescott Fire was knocked into the loser's bracket in the double elimination tournament and stayed on top to the end of the bracket. For the win, they would have to beat their opponent twice. In a strange twist of fate, their opponent was a Central Yavapai Fire District team. The Prescott team successfully defeated their rival firefighters twice and won the coveted prize. Pictured here from left to right is the winning team of J.P. Vicente, Allen "Big Water" Snyder, Tony Reed, Ralph Lucas, and J.D. Fillingim. Prescott ultimately lost at the international level, but Snyder earned his nickname on the trip when facing the big waves of Maui with a mere boogey board.

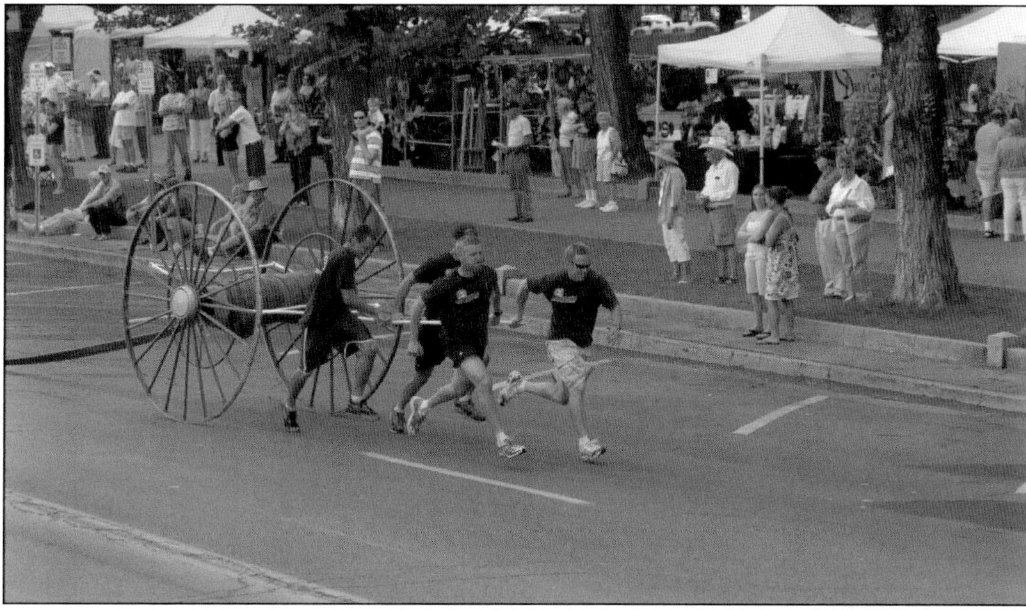

The hose-cart races are still a part of the department's traditions every Fourth of July weekend. The Firefighters Local 3066 runs the event, blocking off Cortez Street at the courthouse for the competition. Prescott, Central Yavapai, Chino, and Mayer Fire Departments are regular attendees, with both men's and women's teams. Pictured here from left to right are Dan Bauman, Brady Brown, Wade Ward, and Travis Epperson. (Courtesy of Wade Ward.)

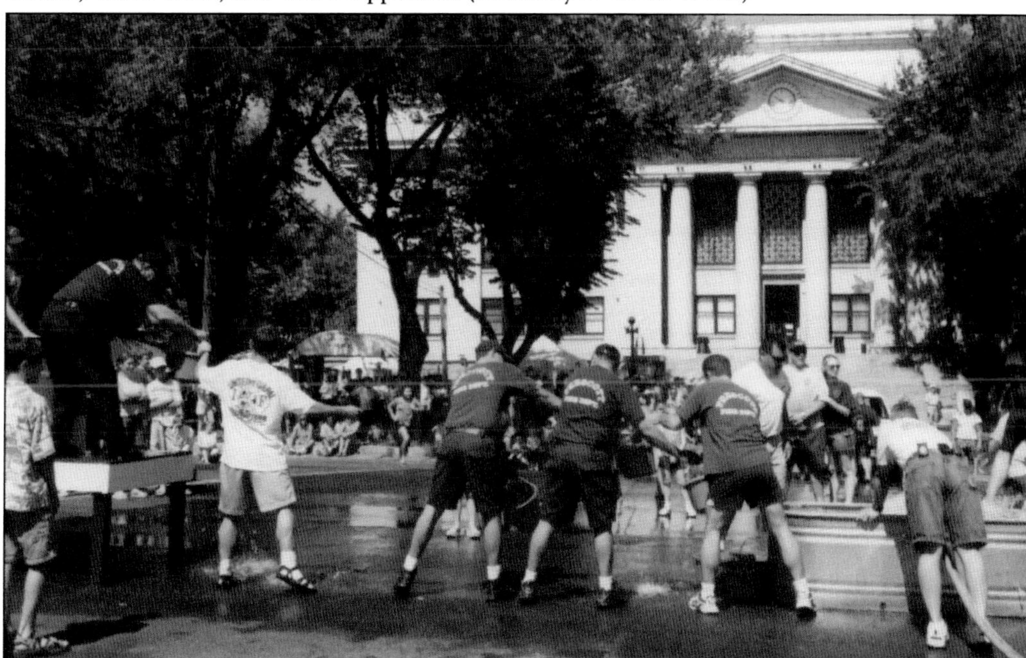

The hose-cart races have also seen additional competitive components. Of late, the bucket brigade race to fill a barrel, pictured above, has been a mainstay, held while the other crews reload the hose carts for the next race. Other intermission races over the years have included hose-rolling contests, barrel shoots, and hose loading.

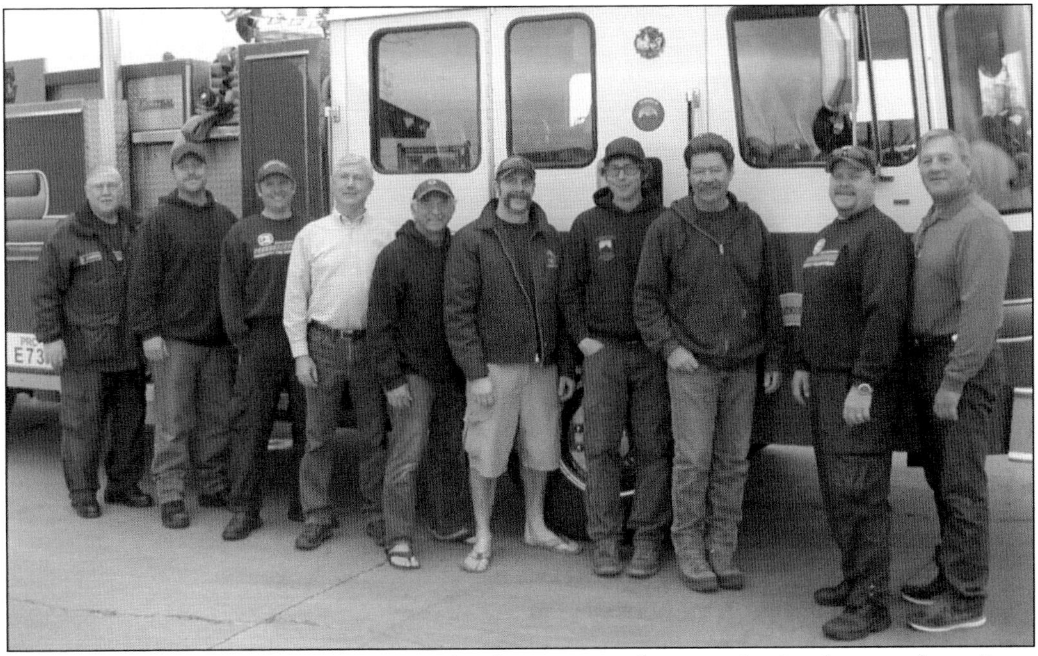

Firefighting for Prescott Fire Department has become a tradition in the families seen here. From left to right are Dan Sanders, Derek Sanders, Aaron Laipple, Paul Laipple, Tim McElwee, Travis McElwee, Tyler Cole, Marty Cole, Troy Steinbrink, and Duane Steinbrink. The father-and-son combinations of Bob and Dean Hinshaw and Darrell and Jared Willis were not available the day of the picture shoot.

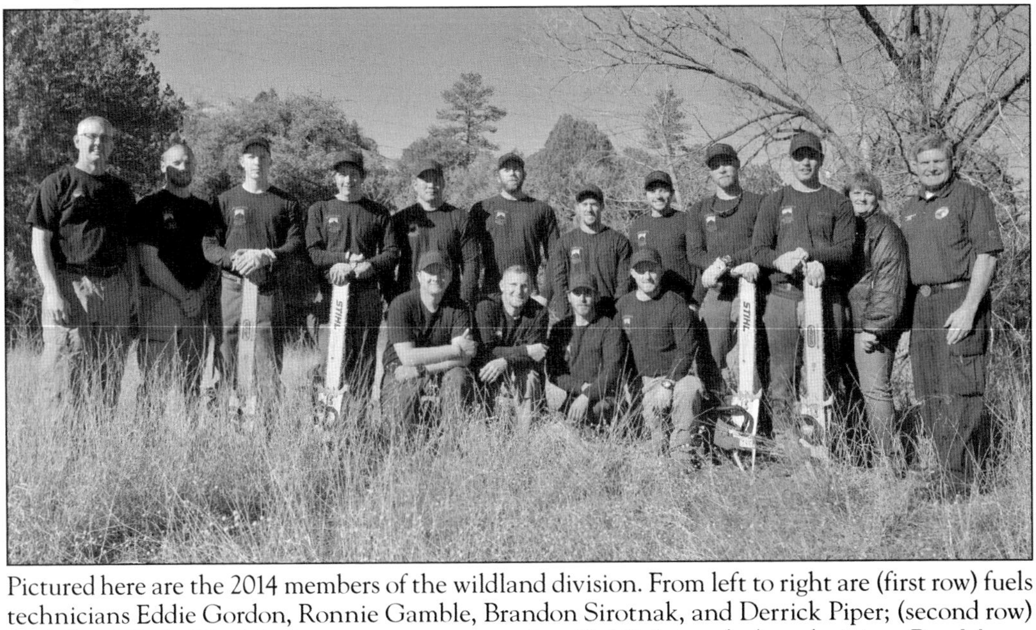

Pictured here are the 2014 members of the wildland division. From left to right are (first row) fuels technicians Eddie Gordon, Ronnie Gamble, Brandon Sirotnak, and Derrick Piper; (second row) firefighter Wade Ward; fuels management supervisor Ted Ralston; fuels technicians Ben Moses, Tyler Cole, Adam Bigalk, Matt Gosney, Randy Justus, Mark Timber, Aaron MacMann, and Shon Do; administrative assistant Masha Collier; and Division Chief Darrell Willis. (Courtesy of Wade Ward.)

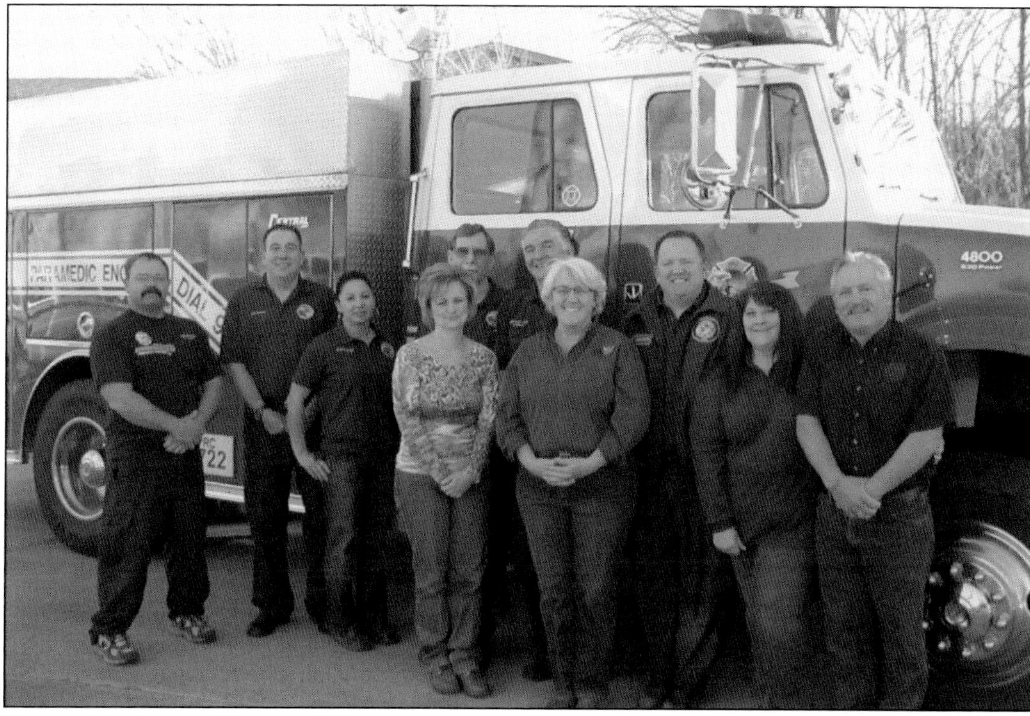

Pictured in front of Engine 722 are a mixture of prevention, administration, and related staff. From left to right are engineer Darrin Osborne, Fire Inspector Len DeJoria, prevention secretary Tiffany Guoladdle, business manager Jaimie Sventek, Fire Inspector Dave Mecca, Division Chief Dave Essex, Arizona Wildfire Academy Assistant Coordinator Libby Reiman, interim-chief Eric Kriwer, training division secretary Karen Cannizzaro, and Arizona Wildfire Academy Coordinator Tony Sciacca. Not pictured are secretary Theresa Morris and Battalion Chief Cory Moser.

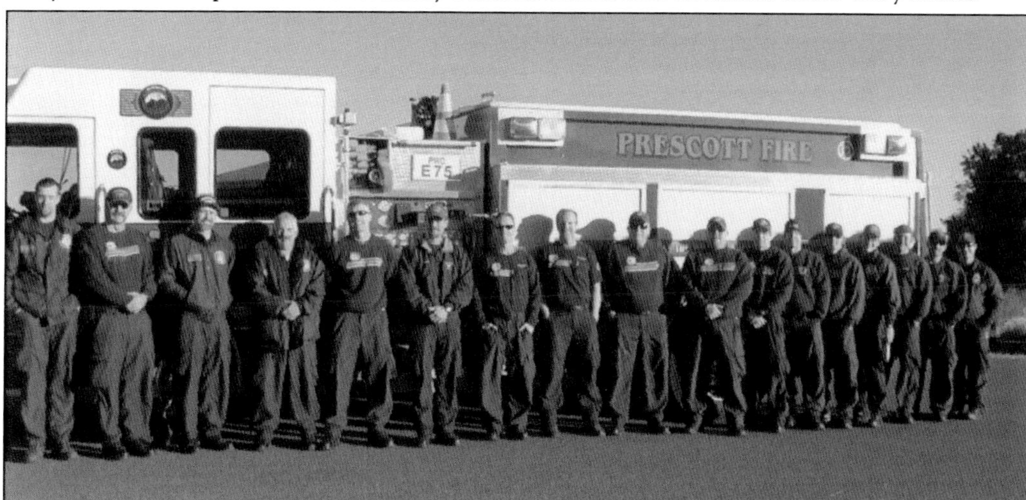

A-shift includes, from left to right, firefighter Doug Harwood; engineer Randy Stazenski; Capt. Kevin Keith; Battalion Chief Don Devendorf; firefighter Ken Webb; engineers Brady Brown and Travis Epperson; Capt. Jason Wallace; Capt. Dan Hutchison; firefighters Troy Steinbrink, Matt DeMenna, Butch Griskowitz, Pat McCarty, and Jeff Archer; engineer Aaron Laipple, Capt. Scott Luedeman, and engineer Nate Seets. Not pictured from A-shift is Capt. Dan Bates.

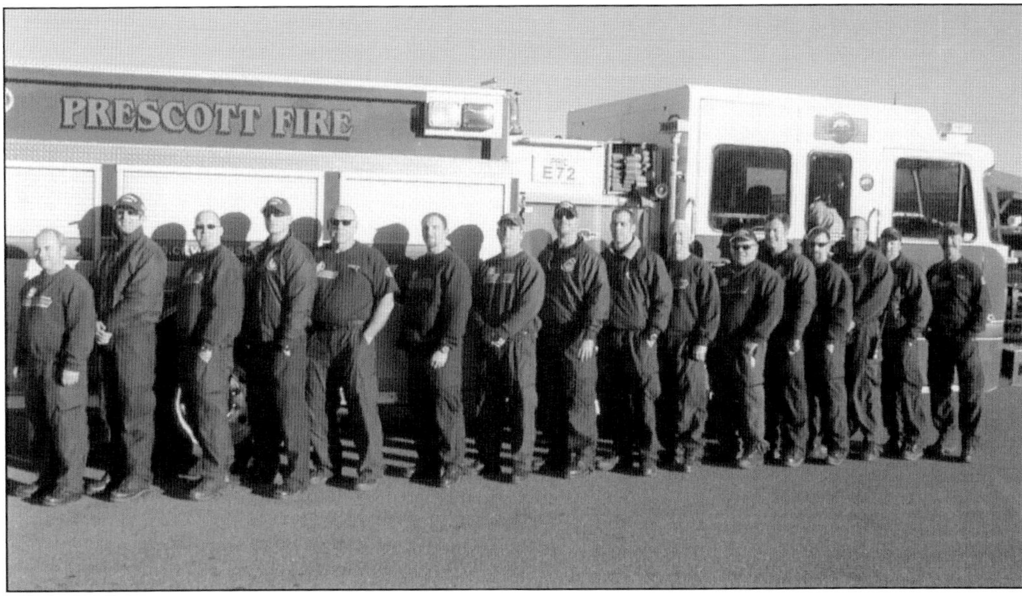

B-shift, shown here in front of Engine 75, includes, from left to right, engineer Ron Kayser; firefighter Adam Julien; engineer Matt Fifield; firefighter David Haskell; Capt. Tim Herrington; firefighter Jordan Pluimer; engineer Brian Padilla-Melton; firefighters Derek Sanders, Phil Baziuk, and Jeff Jones; Capt. Jim Kennedy; firefighters Lance Mills and Jeremy Sarge, Capt. Jeff Moffitt; firefighter Tom Cooley; and Capt. Les Roberts. Not pictured from B-shift are Capt. Dave Peterson, engineer Mark Diedrick, and engineer Mike Lavers.

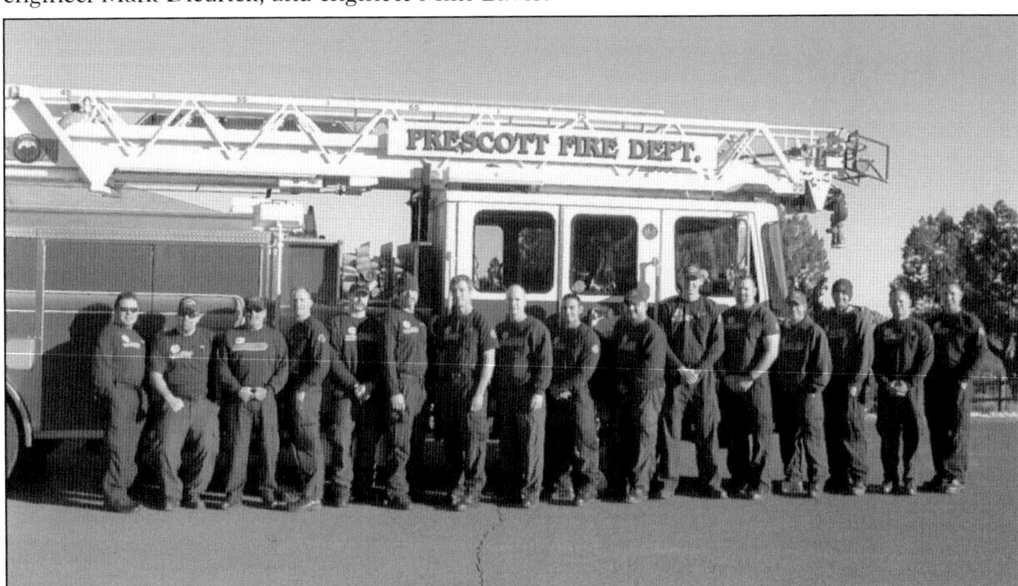

In this C-shift photograph, standing in front of Truck 72 are, from left to right, Capt. J.P. Vicente, engineers Bill Hickey and Tim Miller, firefighter Conrad Jackson, engineer Travis McElwee, firefighter Jason Beyea, Captains Ryan Murphy and Dan Morgan, firefighter Anthony Benitez, engineer Allen Snyder, firefighters Austin Carmen and Matt Reese, Capt. Chad Dougan, engineer Dan Bauman, Capt. James Baker, and Battalion Chief Ralph Lucas. Missing from this C-shift photograph are engineer Dean Hinshaw and firefighter Mark Matthews.

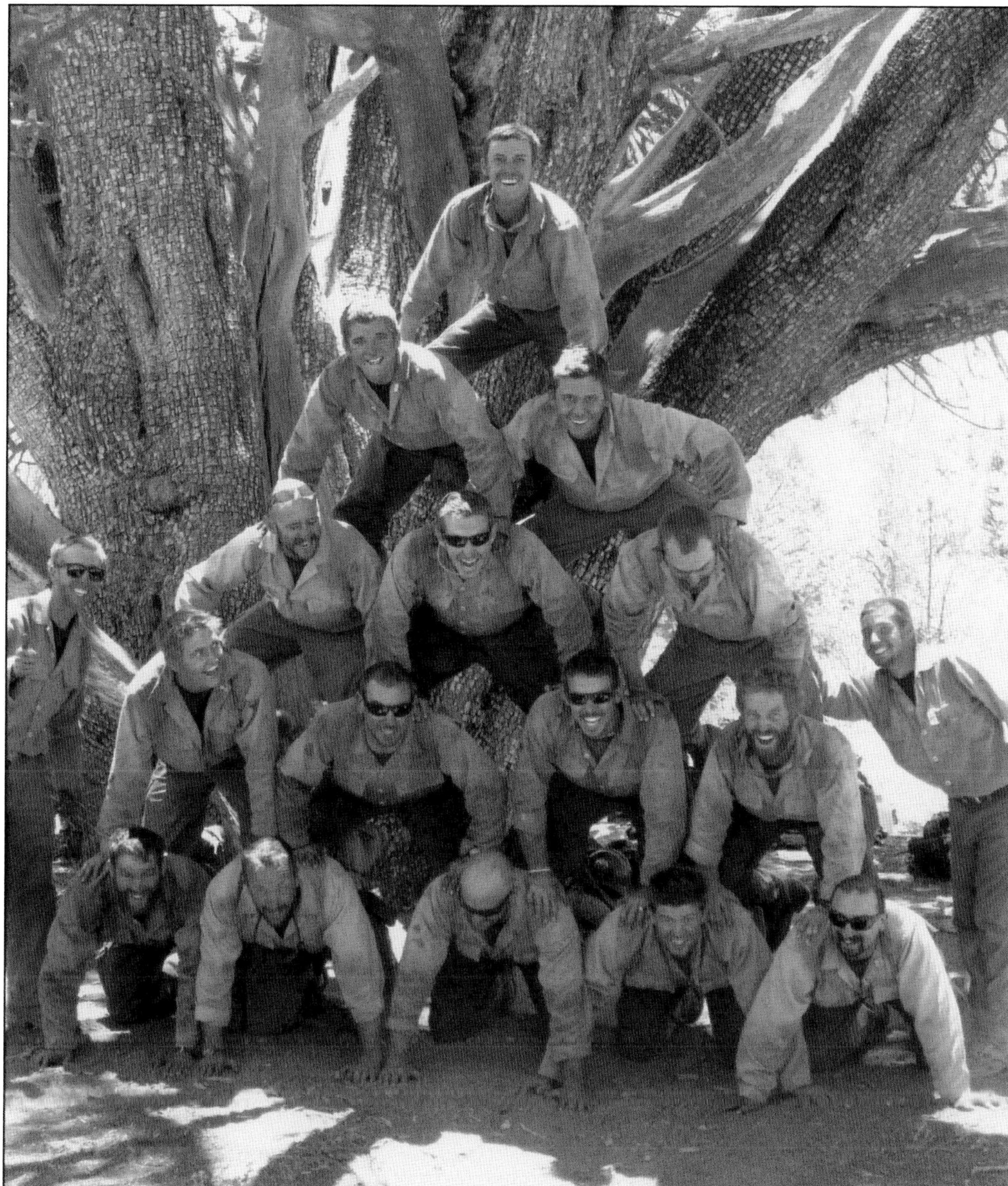

The Granite Mountain Hotshots saved this ancient, record-setting juniper from being destroyed on their namesake mountain the week prior to their deaths. This photograph shows their unity and camaraderie. They represented one quarter of the Prescott Fire Department's personnel and their loss affected not only their own families, but the Prescott Fire Department family and the community they served. They are sorely missed. The fallen firefighters, from left to right, (first row) Dustin Deford, Jesse Steed, Travis Carter, Travis Turbyfill, and Clayton Whitted; (second row) Wade Parker, Joe Thurston, Andrew Ashcraft, and Garret Zuppiger; (third row) Robert Caldwell, Scott Norris, and Sean Misner; (fourth row) Grant McKee and William Warneke. Kevin Woyjeck tops the pyramid with survivor Brendan "Donut" McDonough standing on the left and Anthony Rose standing to the right. The photograph was taken by Chris MacKenzie. Not pictured are Eric Marsh and John Percin. (Courtesy of the MacKenzie family.)

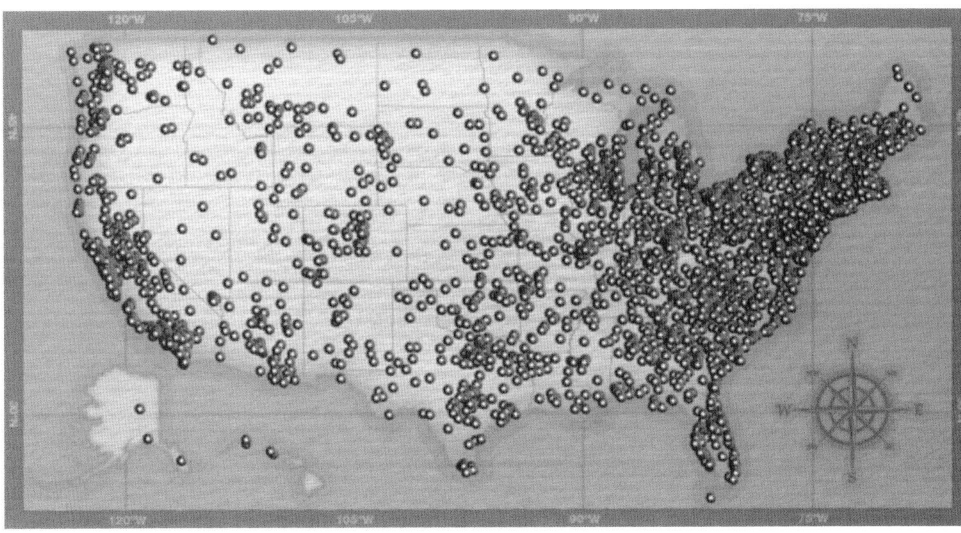